Energy Recovery from Municipal Solid Waste by Thermal Conversion Technologies

Energy Recovery from Municipal Solid Waste by Thermal Conversion Technologies

P. Jayarama Reddy

Renewable Energy and Environment Professional,
former Professor of Physics & Vice-chancellor,
Sri Venkateswara University

CRC Press
Taylor & Francis Group
Boca Raton London New York

CRC Press is an imprint of the
Taylor & Francis Group, an **informa** business

A BALKEMA BOOK

Cover photo: copyright AEB Amsterdam, by Lex de Lang

CRC Press
Taylor & Francis Group
6000 Broken Sound Parkway NW, Suite 300
Boca Raton, FL 33487-2742

First issued in paperback 2019

© 2016 by Taylor & Francis Group, LLC
CRC Press is an imprint of Taylor & Francis Group, an Informa business

Typeset by MPS Limited, Chennai, India

No claim to original U.S. Government works

ISBN-13: 978-1-138-02955-2 (hbk)
ISBN-13: 978-1-138-61211-2 (pbk)

Library of Congress Cataloging-in-Publication Data

Visit the Taylor & Francis Web site at
http://www.taylorandfrancis.com

and the CRC Press Web site at
http://www.crcpress.com

Dedicated to
SAI
&
Professor Arthur L Ruoff,
Cornell University

Table of contents

Preface

Human activities create waste, and the way these wastes are collected, stored, handled and disposed may pose risks to the environment and to public health. The waste sources can be municipal solid waste (MSW), waste from hospitals and industry, the sludge from effluent treatment, and so on. Inapt solid waste management significantly contributes to environmental pollution and to climate change.

The global MSW generation levels in 2012 were approximately 1.3 billion tonnes/year, expected to increase to about 2.2 billion tonnes/year by 2025 posing a serious challenge to the waste managers (World Bank Report 2012). The MSW generation rates are generally influenced by economic development, the level of industrialization, people's life styles, and local climate. The observed tendency has been, higher the economic development (indicated by GDP), higher is the amount of waste generated. The rapid population increase, the fast urbanization growth and the expanding economies of countries, especially developing ones, are contributing to increased rates of MSW generation. Appropriately managed, the waste has the opportunity and potential to become an asset – a source of clean energy.

Energy-from-Waste or Waste-to-Energy technologies, as commonly referred to, are the treatment processes that produce energy from a waste source in the form of heat, electricity, transport fuels or high-value chemicals. Power can be produced and distributed through the grid systems. Heat can be generated both at high and low temperatures and distributed for district heating or utilized for specific processes. Fuels that can replace fossil fuels can be extracted from certain types of waste. The most focus is on MSW which is increasingly produced, though medical and hazardous wastes and non-recyclable plastics are also treated to a large extent.

However, an increasingly tough environmental, economic and technical factors have slowed down the growth of these technologies. Even though these are well developed, the variations in the composition of MSW, the complexity of the design of the processing facilities, and the air-polluting emissions particularly toxic gases were always the issues for debate.

Efforts from several different perspectives are required for the development and success of these projects because the social, economic and environmental issues become crucial in the decision making process of the technology. Moreover, the scenario of solid waste in terms of generation rates, composition and treatment is dramatically changing due to growing population, increasing urbanization and economic advancement.

Although most of the MSW is generated currently in North America and European countries, the fastest growth in waste generation for the coming decade is expected mainly in emerging economies in Asia, Latin America and Africa. A shift towards an increased fraction of plastic and paper in the waste composition that happened in the high-income countries is expected to take place both in the middle- and low-income countries with the growing urbanization and economic enlargement in these countries.

As of now, the most common and well established waste-to-energy technology is Incineration (Combustion), a thermal process, generating heat and power. Many countries world-wide have acknowledged waste incineration as an essential part in an overall sustainable waste management scheme clearly recognizing the disadvantage of landfilling from the point of greenhouse gas emissions and the long-term risk of pollution to soil and ground water.

In the incineration process, the waste is combusted with excess air in a closed chamber under 'controlled' process parameters to release the stored chemical energy. The end products of the process are hot flue gases and solid residues. From the flue gases, heat is recovered and utilized or power is produced before let into the atmosphere after removal of pollutants utilizing flue gas cleaning devices. The solid residues, in the form of bottom ash or slag and fly ash can be treated and reused in the building and construction industries. The 'as-received' waste is processed depending on its quality, composition and the selected incineration system before entering the combustion chamber. The size of the plants can vary significantly, in terms of waste input capacity and of power output. The efficiency for the incineration process (state-of-the-art plants) is typically 20–25% if operating in CHP mode and up to 25–35% if only power is produced.

In the early 1980s, the incineration technology was objected by the public due to the pollutant emissions, especially the toxic dioxins/furans. This has led not only to research efforts to develop systems to decrease the environmental impact of the incineration technology but also more tightened regulations of air emissions by governments. These R&D activities, within almost a decade, have produced the most effective air pollution control (APC) systems that has radically changed the face of waste incineration technology. As a result, by early 2000, the incineration technology reached a status of environmental-friendly in the public perception (Vehlow 2015).

Another most significant outcome of the public opposition against waste incineration technology was the development of advanced thermal conversion processes – Pyrolysis and Gasification. These processes are perceived to have the potential to be more efficient in the recovery of energy and materials from the solid waste and in reducing the pollutant emissions, particularly the toxic ones. Several variants based on these two technologies have been developed and demonstrated; some of them have reached even commercial status. Yet, the commercial experience with these processes is very limited compared to that of incineration technology. Moreover, the data on their economics are not clear. These are mostly used for treating special wastes: sewage sludge, hazardous waste and incineration residues, plastic waste and so on. The variants treating municipal solid waste are generally two-stage processes with the second stage being a combustion process. Hence their main final gaseous product is a flue gas of similar quality as that from waste incineration plants. Therefore, in principle, the plants are equipped with the same APC systems. This product known as synthetic gas (syngas) can be used to generate power or alternately reformed to serve as a source for producing transport fuels and high value chemicals. Early attempts to implement these advanced technologies in Europe have failed, mainly because of economic constraints and partly of technical problems. However, Japan and to some extent South Korea, have fully exploited these two technologies and Japanese companies developed their own pyrolysis and gasification systems which are now being used outside Japan also. The U.S. has renewed interest in these processes in recent years and several firms have come up as technology developers/suppliers.

Regarding the technology-related costs, capital costs can vary significantly depending on the extent of flue gases cleaning and other produced residues. Operation and maintenance costs have a minor impact on the total expenses of the facility and are mainly related to the amount and nature of the waste stream. Despite the global economic crisis, the global market of waste to energy technologies has registered an average annual increase of 5% in the past few years and is expected to reach a market size of US$ 29 billion by 2015 at a 'compounded annual growth rate' of 5.5% (Frost & Sullivan, 2011). The main drivers for this growth could be an increasing waste generation, high energy costs, growing concern for the environment, and limited landfilling space. These energy recovery processes help solve these issues by sinking the waste volumes by 70 to 90% and reducing greenhouse gas emissions. Additionally, the legislative and policy measures, mainly by the European countries (the EU Directives), the U.S (EPA protocols) and Japan have significantly improved the growth of this market and the advanced technology solutions.

Substantial progress has been achieved: almost 2240 waste-to-energy plants are active in the world with a disposal capacity of 270 million tons per year; it is estimated that almost 500 new plants with a capacity of about 150 million annual tons to be installed by 2023 (Mark Doing, www.ecoprog.com, 2015). However, more research efforts are required focusing on finding complete solutions for major reduction of pollutant formation, and for ensuring creation of less but high quality solid residues, lower stack emissions and high efficient power generation burning the total municipal solid waste.

This book discusses these three technologies – Incineration, Pyrolysis, Gasification – and the formation of pollutants in the product gases and their control/cleaning methods in a fairly good detail in Chapters 2, 3 and 4. Chapter 1 is devoted to a general brief on global waste generation and its projection, waste hierarchy followed, introduction to waste thermal conversion technologies and so on. Chapter 5 comprises Case studies: the waste-to-energy plants *in operation,* all of them based on Incineration technology, except one. A few commercial gasification (or combination with pyrolysis/combustion) technologies operating successfully are outlined in Chapter 4. These case studies are intended to demonstrate confidence in the potential of these technologies.

Plastic waste generation has been alarmingly increasing in recent years because of their use in numerous applications/sectors. Hence handling of plastic waste especially non-recyclable plastics has assumed significance. Considerable amount of research has been done in this area. Laboratory-scale studies are carried out in several institutions to generate power or to extract oils, and most of them have the potential to become commercialized. A few basics relate to plastic waste treatment, particularly energy recovery in terms of oil and solid char that can be utilized to produce transport fuels and chemicals are covered in Chapter 4.

Several books in this area, well written with different perspectives are available; yet, it is author's objective to comprehensively present the well-developed and more widely implemented 'waste thermal conversion technologies' to the level of an undergrad student or any one inquisitive to know the progress in the field. The book is also intended to serve as a handy guide to a consultant or a policy maker or those engaged in the waste-to-energy area. Hence, mathematical treatments and finer engineering details are not included, and emphasis is laid on the principles, instrumentation, operation and performance. Nonetheless, considerable references are provided in each chapter to dig for more details. The book and the references may serve as a useful tool particularly to a student aspiring to pursue further in the field. Four 'Annexures' are added to include related aspects along with references.

The economics of these technologies are not discussed, though several reports have presented. Apart from the technology chosen, the costs are influenced by local factors related to waste management practices, country's regulations and policies concerning energy generation and environmental safety, among others. These factors differ significantly and the cost analysis, in the author's view, may not hold for all the places and for all the time. However, broad guidelines to evaluate the economics of a waste-to-energy plant at any location is given in an annexure.

Nevertheless, one thing is clear which driven the author to attempt this comprehensive book: the recovery of energy contained in the MSW is getting more attention globally as MSW is recognized as a source of renewable energy, and not simply a waste to be wasted. This recognition will undeniably enhance the energy security at least in urban areas, and the environmental safety by minimizing greenhouse gas emissions that cause global warming.

Foreword

The world is awash in waste due to the current management practices that do not fully recognize the inherent value contained in municipal waste streams. To put this in perspective, as of 2014 nearly 11% of the world population lives without access to clean water or food and 18% without electricity. However 100% of the population has access to waste; because each and every person produces some amount of waste. In fact World Bank data shows that across a median income ($US year^{-1}) range from 5,000 to nearly 75,000 waste generation rates do not go below approximately 2 lb person^{-1} year^{-1}. This begs the question; is there a minimum amount of waste generation the human population must produce to sustain itself?

Before answering this question, a more immediate concern is how to manage the huge volumes of waste already generated. The waste hierarchy dictates that the most desirable action is to reduce the amount of waste generated. That is stop the problem at the source. If that cannot be done, the next best option is to reuse the material, i.e. do not process, just repurpose. If that cannot be achieved, then recycle as much material as possible. Yet it is well known that recycling has limits both technically and economically. Once all the reduction, reuse and recycling are accomplished, the immediate next option to extract energy from the waste stream. The extraction of energy can be in the form of heat or power. Furthest down on the hierarchy is the disposal of material in sanitary landfills. While following the directives of the hierarchy must be respected, unfortunately the only two options that can safely manage the waste commensurate with the scale that it is produced are the two lowest; energy recovery and landfill.

This text by Dr. Jayarama Reddy details the myriad technologies developed over the years to thermally process municipal solid waste (MSW) streams for power generation. It is an excellent assembly of the specifics of the numerous technologies developed and deployed for thermal conversion of MSW for power applications ranging from traditional combustion to advanced plasma and pyrolysis. Importantly, Dr. Reddy articulates a great number of hard to find, yet essential pieces of information associated with specific systems and conversion systems. This text is invaluable to those who want to learn about exactly why a technology was developed and its performance when finally deployed at a relevant scale. It is clear Dr. Reddy has an intimate knowledge of the specifics and details behind why certain technologies succeed and why others fail. The realm of thermal conversion of MSW technology development has been a continuing saga over the past four decades. This has spanned from redesigning legacy systems to exploring the most advanced way to convert MSW to power and energy. The amount of particular detail and focused discussion on each technology class and process demonstrates the tremendous insight Dr. Reddy has in this field. Engineers, designers, inventors and students would do well to closely examine and absorb the information provided to fully understand the intricacies and nuances associated with such a challenging subject as the thermal conversion of MSW for power applications. The information provided here should serve as a study guide for those interested in developing such systems to avoid the failures of the past.

Professor Marco J. Castaldi, Ph.D.
Department of Chemical Engineering &
Director, Earth Engineering Center,
The City College, The City University of New York, New York, NY

Acknowledgements

At the outset, I express my sincere gratitude to Professor Dr. Marco Castaldi, Department of Chemical Engineering and Director, Earth Engineering Center, The City College, City University of New York, New York, for the expert reviewing of the manuscript and writing Foreword. In spite of heavy professional schedules, he could complete the work admirably much earlier than expected. I'm personally obliged to him for his time, effort and splendid support.

I'm highly thankful to Professor Yiannis Levendis, College of Engineering Distinguished Professor, Department of Mechanical & Industrial Engineering, Northeastern University, Boston, for his help and suggestions during the preparation of the manuscript; Prof. Dr.-Ing. habil. Klaus Goerner (Universitat Duisburg-Essen), Prof. Dr.-Ing. Rudi Karpf (Ingenieurgesellschaft fur Energie), Dr.rer.nat. Juergen Vehlow & Dr. Hans Hunsinger (Karlsruhe Institute of Technology), Dr. Joachim Werther (Hamburg University), Prof. Qunxing Huang (Zheijiang University), Dr. Tony Murphy & Dr. Wes Stein (CSIRO, Australia), Prof. K. Yoshikawa (Tokyo Institute of Technology), Prof. Carlo Vandecasteele (University of Leuven), Dr. Y. Kodera (AIST, Tsukuba, Japan), Ryo Makishi (Nippon Steel), Francis Campbell (President, Interstate Waste Technologies), Dr. Michael Keunecke (Hitachi Zosen Inova AG), Dr. Juan Unda (SENER2 – Zabalgarbi plant), Dr. Bary Wilson (Enviropower), Ross Patten (Agilyx Corp.), Westinghouse Plasma Corporation (division of Alter NRG Corp.), Gershman, Brickner & Bratton, Inc., European IPPC Bureau, SINTEF Energi AS, American Chemical Society, EBARA Environmental Plant Co. Ltd., and JFE Engineering (S) Pt Ltd., and many others for permitting to use their research results and figures; my grand-daughter Hitha for preparing a few figures; and Prof. N.J. Themelis (Columbia University), Prof. P.T. Williams (University of Leeds), and Professor Jose Aguado (Universidad Rey Juan Carlos) and their co-workers whose publications were very valuable in the preparation of the text.

I extend my hearty thanks to my cherished friends and family, particularly, Jagan and Vijaya, Sreeni and Geeta, my wife for their understanding and support, my grand-kids, Hitha, Tanvi, Diya and Divi for their cheer; and the editorial staff at CRC Press, particularly Alistair Bright and Richard Gundel for their superb support.

P. Jayarama Reddy

Chapter 1

Waste generation & management

1.1 INTRODUCTION

Municipal solid waste management (MSWM) has been a complex challenge to the civil society. Human activities create waste, and the way these wastes are collected, stored, handled and disposed may pose risks to the environment and to public health. Improper solid waste management has contributed greatly to air, land and water pollution, and to climate change.

Managing solid waste by the municipalities or the local bodies is an exhaustive service. They need capacities in procurement, contract management, professional and unionized labor management, and expertise in investments, budgeting and finance. More importantly, MSWM requires a strong social bond between the municipality and the community. A successful treatment of MSW should be safe and environment-friendly (Sakai et al. 1996).

In recent years, waste generation rates have been growing rapidly worldwide. The alarming rate of solid waste generation trends in any country can be seen parallel to population increase, growth in urbanization and industrialization, affluence, and growth-centric economy. This is particularly the case in developing countries and the environmental burden continues to be a major issue for them (e.g., Jayarama Reddy 2014a). Comparison of conditions related to MSW management in developed and developing countries brings indicators that quantify the problem.

Open or organized dumping or landfill or burning with no energy recovery, and/or composting have been the traditional way of disposing municipal solid waste (e.g., Jayarama Reddy 2011, 2014a). Recent trend has been 'Integrated Solid Waste Management (ISWM)', a comprehensive program consisting of waste prevention, recycling, composting, and disposal. An effective ISWM system considers how to prevent, recycle, and manage solid waste in ways that most effectively protect human health and the environment. ISWM involves evaluating local needs and conditions, and then selecting and combining the most appropriate waste management activities for those conditions (Kokalj and Semac 2013).

1.2 WASTE GENERATION

The global MSW generation levels in 2012 are approximately 1.3 billion tonnes per year, and are expected to increase to about 2.2 billion tonnes per year by 2025 posing

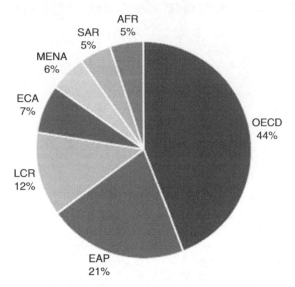

AFR: Africa, EAP: East Asia Pacific; ECA: Europe and Central Asia;
LCR: Latin America and Caribbean Region; MENA: Middle East and
North Africa; SAR: South Africa Region; OECD: Organization for
Economic Cooperation and Development

Figure 1.1 Waste generation by region (source: World Bank Report 2012).

a serious challenge to the waste managers (World Bank Report 2012). This represents a significant increase in per capita waste generation rates, from 1.2 kg/capita/day in 2012 to 1.42 kg/capita/day in 2025.

The current generation of MSW by region in the world is shown in Figure 1.1. Figure 1.2 shows the waste generation in the world countries grouped as lower income, lower middle income, upper middle income, and high income. As the income level of a country increases, the waste generation per capita per year increases, from 219 kg/capita/year for lower income country to 777 kg/capita/year for high income country. The projected waste generations to 2025 are also shown in the figure.

In East Asia and the Pacific Region, the annual waste generation is approximately 270 million tonnes per year. This quantity is mainly influenced by waste generation in China, which makes up 70% of the regional total; and the per capita waste generation ranges from 0.44 to 4.3 kg/capita/day for the region with an average of 0.95 kg/capita/day.

In Eastern and Central Asia, the waste generated per year is at least 93 million tonnes. Eight countries in this region have no available data on waste generation in the literature. The per capita waste generation ranges from 0.29 to 2.1 kg/capita/day, with an average of 1.1 kg/capita/day.

In South Asia, approximately 70 million tonnes of waste is generated per year, with per capita values ranging from 0.12 to 0.51 kg/capita/day and an average of 0.45 kg/capita/day (The WB Report 2012).

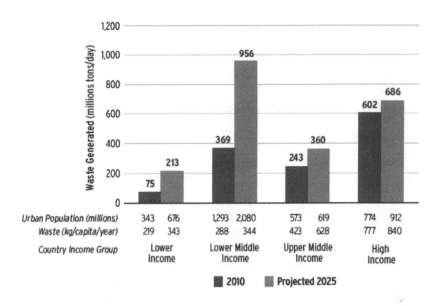

Urban Population (millions)	343	676	1,293	2,080	573	619	774	912
Waste (kg/capita/year)	219	343	288	344	423	628	777	840
Country Income Group	Lower Income		Lower Middle Income		Upper Middle Income		High Income	

■ 2010   ▨ Projected 2025

Figure 1.2 Urban waste generation by income level and year (Source: World Bank Report 2012).

Table 1.1 Urban waste generation: Current data and projections to 2025.

Region	Current data (urban waste) Per capita (kg/capita/day)	Total (tons/day)	Projections for 2025 (urban waste) Per capita (kg/capita/day)	Total (tons/day)
AFR	0.65	169,119	0.85	441,840
EAP	0.95	738,958	1.5	1,865,379
ECA	1.1	254,389	1.5	354,810
LCR	1.1	437,545	1.6	728,392
MENA	1.1	173,545	1.43	369,320
OECD	2.2	1,566,286	2.1	1,742,417
SAR	0.45	192,410	0.77	567,545
Total	1.2	3,532,252	1.4	6,069,703

AFR: Africa, EAP: East Asia Pacific; ECA: Europe and Central Asia; LCR: Latin America and Caribbean Region; MENA: Middle East and North Africa; SAR: South Africa Region; OECD: Organization for Economic Cooperation and Development. (Source: The World Bank Report 2012).

In Low income countries, the situation is much different: the per capita waste generation is low (around 0.3–0.4 kg/d). In these countries, population increase, urbanization and increased consumerism are predicted as future trends which may result in increase in solid waste generation. The economies have grown in these countries over past decade. Across all developing countries the growth trend is approximately 6–7%. e.g., Sub Saharan African Economies have grown nearly 5% per year.

In Table 1.1 are given the estimates of waste generation for the year 2025 as expected according to current trends in population growth in each region.

The United States is one of the countries with highest waste generation. In 2013, Americans generated about 254 million tons (U.S. short tons) of trash and recycled and composted over 87 million tons of this material, equivalent to a 34.3% recycling rate. On average, Americans recycled and composted around 0.69 kg/capita/day out of the individual waste generation rate of 2 kg/capita/day. The U.S. (EPA) is transitioning from focusing on waste management to focusing on Sustainable Materials Management (SMM) which refers to the use and reuse of materials in the most productive and sustainable way across their entire life cycle. SMM conserves resources, reduces waste and minimizes the environmental impacts of the materials used (US EPA 2015). In 2009, the European Union produced 283 million tons of MSW at 1.41 kg/capita/day with noticeable variations from country to country. For example, while Poland and the Czech Republic had per capita MSW generation rates on the low side at around 0.86 kg/capita/day, Denmark had generation rates on the high side of around 2.26 kg/capita/day. The United Kingdom and Germany had MSW generation rates around the E.U. average at 1.45 and 1.59 kg/capita/day, respectively (European Commission 2008).

Interestingly, the top ten MSW producing countries include four developing countries – Brazil, China, India, and Mexico – because of the large size of their urban populations as well as due to prosperity and high-consumption lifestyles of the city dwellers. Even among the top ten, there is a wide disparity in output, with USA generating nearly seven times more urban refuse than France does which is in tenth place (WMW 2012). Similar to urbanization and prosperity growth rates, MSW growth rates are fastest in China, parts of East Asia, Eastern Europe and the Middle East (Hanrahan et al. 2006). Further, the production of cheap consumer goods on a massive scale to bring down the market prices generally sacrifices the quality and life of goods. People are therefore inclined to simply throw away and go for new items instead of servicing or maintaining the old ones.

In an era of limitless business ingenuity but limited resources, the sustainable management of natural resources is increasingly being debated about how to achieve economic growth without compromising the environment and the human health upon which that growth depends. By looking across the life cycle, businesses can find opportunities that enhance and sustain their value scheme and reduce risk through sustainably managing materials (US EPA 2015).

According to the UN Environment Programme (UNEP), "Humans are consuming resources and producing waste at a greater scale than ever before and per capita consumption levels are projected to increase with continued development." For every 1% increase in GDP, resource use has risen 0.4%. Data indicate that global material resource use during the 20th century rose at about twice the rate of population. The growth rate in materials use was still lower than the pace of growth of the world economy. Nevertheless, resource use is still on a steep rise and the demands in the future may be still higher, given projections around future world population growth, economic growth and energy and material consumption. According to the World Resources Institute, "one half to three quarters of annual resource inputs to industrial economies is returned to the environment as wastes within just one year" (USEPA 2015).

1.3   MSW MANAGEMENT SYSTEM

1.3.1   Waste hierarchy

An efficient management system for collection and disposal of solid waste is funda-mental for any community. Generally, the demands on the solid waste management system increase with the size of the community and its per capita income.

Waste management is a complex issue driven by several factors: regional and seasonal waste composition, disposal and recovery technologies, local and national policies, economic factors such as the price of energy or the cost of landfill and so on. Land is very expensive in urban areas to establish landfill, and its location is often opposed by neighboring residents for fear of environmental pollution and bad odors. Waste burning with no energy recovery was practiced to reduce the volumes that go to landfills.

The U.S. Environmental Protection Agency has developed a waste manage-ment hierarchy defining its regulatory policy preference for solid waste management (Figure 1.3).

The International Solid Waste Association states: "... the waste hierarchy is a valuable conceptual and political prioritisation tool which can assist in developing waste management strategies aimed at limiting resource consumption and protecting the environment" (ISWA 2009). Hence, priority is given in order to waste minimisation, reuse, recycling, waste-to-energy, and finally landfill.

The waste-management sector worldwide faces the following important issues: (a) achieving far greater reduction of waste quantity at the source; (b) improving compilation of waste statistics to measure waste trends; (c) increasing recycling efforts to make a significant impact on reducing landfill and the pollution caused by waste, and developing a viable commercial infra-structure for more successful recycling, and (e) enhancing energy recovery activity (heat and electricity) from waste (WMW 2012).

It was suggested that MSW management be integrated into a materials manage-ment approach, a 'circular economy', involving a series of policies to reduce the usage of some materials and to reclaim or recycle most of the rest. Japan made the circu-lar economy a national priority since the early 1990s by a steady expansion of waste reduction laws and achieved remarkable success. Resource productivity is rising to reach more than double by 2015 from 1990 levels. Besides, the recycling rate is pro-jected to roughly double over the same period and the total material sent to landfill to decrease to about a fifth of the 1990 level by 2015 (WMW 2012).

Waste prevention and minimization effectively reduces usage of limited resources, and also avoids environmental impacts associated with waste handling, waste treat-ment and waste disposal activities. The resource recovery and recycling reflect that solid wastes are materials and by-products with no value for the keeper. But, what may be considered 'waste' may change with the circumstances of the possessor as well as in time and place. What is considered as waste in an affluent society may be useful in a poor society for reuse. Waste may be converted into a resource by moving to a new place or through proper treatment. Such a conversion depends on the costs involved and whether the economy is looked upon as a private business or a national or even a global priority (WB 1999).

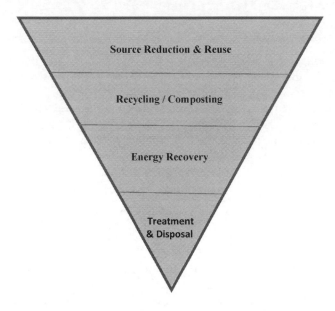

Figure 1.3 Waste Management Hierarchy graphically represents the options of handling waste, from the most desired way of 'source reduction' to the least desired way of 'treatment & disposal' (redrawn from the EPA document, 'MSW in the U.S. – 2011 Facts & Figures, May 2013').

Recycling conserves natural resources, reduces emissions from manufacturing items out of new materials, saves energy, generates revenue and creates manufacturing jobs. Treating waste to recover energy (heat/electricity) involves machinery and huge installations that require large investments and operating costs. Hence, the countries with fewer financial resources, should acquire complete knowledge of the existing system and the type and rate of the waste they generate before introducing energy recovery step. This is quite a challenge, except in a highly organized waste management system (WB 1999). The history of solid waste management with reference to the U.S. was explicitly discussed by Lanny Hickman (2003).

Municipal solid waste management (MSWM) solutions have evolved significantly over time mainly due to the community reaction and country's technological and policy developments. The increasing awareness of the benefits of resource recovery and recycling and energy recovery from waste have led to complex and evolving regulations. Solving one problem often introduces a new one, and if not well executed, the new problem is often of greater cost and complexity (The WB Report 2012).

1.3.2   Recycling & energy recovery

The average city's municipal waste stream is made up of millions of separate waste items. In many cases, a city's waste stream is originated from numerous factories and manufacturers that produce goods for other countries. Some of the waste fractions are large in volume; waste fractions such as food and organic material and paper are easier to manage, and wastes such as multi-laminates, hazardous and e-wastes create unduly big problems. Hybrid and voluntary take-back programs are recently emerging; however, these programs are yet to function efficiently.

Notwithstanding these differences in MSW generation and nature, *recycling* has been pursued in the last century by several countries – North America, European countries, Japan and a few others. Source-separated recycling, where recyclables such as paper, plastic, and glass are placed in separate bins for collection, is widely used in developed countries. In contrast, single-stream recycling has been the practice in many developing countries except very few; all recyclables are mixed and collected from a single bin and sorted at a materials recovery facility (e.g., Jayarama Reddy 2014b). The variations in these end-of-life solutions for waste depend very much on social and policy drivers related to landfill, composting, recycling, and energy recovery.

In the U.S. 34.5% was recovered through recycling and composting, 53.8% was landfilled and 11.7% was combusted for energy recovery in 2013. On average, Americans recycled and composted 0.69 kg out of the individual waste generation rate of 2 kg/person/day (USEPA 2015). The European Union as a whole has a higher recycling (including composting) and energy recovery rate at 40.6% and 16.1% respectively, compared to U.S. The recycling and energy recovery rates in EU vary considerably by country, with Germany and Austria recycling or composting 63.6% and 69.9% of their MSW. Recycling rates in the UK rose faster from just 12% in 2001 to 39% in 2010 on par with average value of EU. Countries such as Germany, Austria and Belgium, already recycle more than 50% of their municipal waste. In some countries, the recycling rates are far less or have fallen: for example, Greece recycles 18% only, up from 9% in 2001 and Romania recycles just 1%; in Norway recycling rate has fallen from 44% to 42%, and Finland from 34% to 33% (Adam Vaughan 2013).

In contrast, in the low income countries, the landfill dominates among management practices (in some regions alongside informal disposal open dumps and open burning). As the composition and volumes of municipal waste are expected to increase in these countries due to population growth, a need to improve the management structure, investments flow and the plan for treatment of waste is imminent.

In the most populous countries, India and China, the MSW practices are relatively better among the developing countries. In 2012, Indian waste generation amounted to 0.2–0.87 t/capita/yr (around 0.55–0.65 in urban areas) while China had a generation rate of 0.98 t/capita/yr (Planning Commission TF report 2014). In India, ordinary landfill and composting to an extent are the common disposal methods, but landfill construction has become expensive in big urban areas due to rise in costs for land. Therefore uncontrolled dumping and/or burning of waste even in public places still continue. The absence of segregation at source makes recycling activity more difficult. At the dump sites, unorganized rag pickers estimated around one million in number, collect paper, plastics, metal scrap, glass bottles etc as they have economic value (see Jayarama Reddy 2011, 2014b). Therefore, the waste received by the waste treatment companies is that left behind by these collectors, and hence generally of very low content quality. The putrescible content remains high in the waste which has a calorific value of 6.8–9.8 MJ/kg. Thus, energy recovery technologies available in the market (mostly applicable to European countries and North America) are inappropriate for this type of waste.

In China, MSW growth is fastest. The calorific value of the waste is highly variable, ranging from 4 to 11 MJ/kg; and the moisture content and ash content are also high.

Table 1.2 Characteristics of MSW in low, medium, and high GDP countries.

Example country	*Low-GDP countries* India	*Medium-GDP countries* Argentina	*High GDP countries* EU-15
GDP US$/capita/year	<$5,000	$5,000–$15,000	>$20,000
MSW kg/capita/year	150–250	250–550	350–750
MSW collection rate	<70%	70%–95%	>95%
% putrescible waste in MSW	50%–80%	20%–65%	20%–40%
Heating value kcal/kg	800–1,100	1,100–1,300	1,500–2,700

(source: Lacoste and Chalmin, 2006).

A 2005 World Bank report noted: 'Shanghai, which has the most 'internationally standard' waste stream (i.e. higher fraction of plastics and papers and less moisture) still has a waste composition that is barely autogenic (a high enough heating value to burn on its own) (World Bank 2005). As per 2008 data, only 55% of waste is treated; of which 80% land filled, and 15.2% incinerated (697 t/day). As the landfill space is scarce in big cities, the energy recovery from waste activity has been on the rise since late 1990s (e.g., Cheng and Hu 2010, Zhou and Chen 2012). Table 1.2 shows the average characteristics of MSW in low-, medium- and high GDP countries for comparison.

Japan, contrary to the situation in China and India, has high energy recovery activity because of scarce land resource. Of the total waste generated, 54.4 mt in 2003, nearly 74% was incinerated, probably the highest in the world. The landfill fraction is very low at 2%.

Some of the developing countries in Asia and Africa face a peculiar problem of waste being dumped from outside. Due to increasing landfill charges and tough legislation demanding organisations and businesses to recycle more, most of the waste generated in EU (particularly The Netherlands, Switzerland, Belgium and Germany) is sent to West Africa and Asia, mainly, Ghana, Nigeria, India, Pakistan and China, under the guise of 'used goods'. The handling of this stuff is done in most disorganised manner creating air pollution as well as health problems to the unskilled workers. More than a third of the waste paper and plastic collected in UK is sent to China for recycling. Recyclable plastics sorted from the municipal waste, estimated at 12 m tonnes/year, flows from affluent countries to Asia, especially China. Among the plastic scrap exporters, a few European countries including Germany, Hong Kong, the U.S., Japan and the UK represent the top ones (Moses 2013). More details on recycling and reuse worldwide is available in the literature.

1.3.3   Markets for secondary materials

The growing vagaries of secondary materials markets all over the globe have emerged as an additional challenge in the last ten to twenty years. Marketing of secondary-materials derived from the waste has become a global business. For example, the price per tonne of waste paper in New York City is often based on the purchase price in China; the majority of waste recycled in Buenos Aires is shipped to China, and so on

(WB Report 2012). The wavering secondary materials prices have made planning more difficult. The price is often predictive of economic trends, significantly dropping during economic recessions. There are some hedging opportunities for materials pricing; however, secondary materials marketing does not have the same degree of sophistication as other commodities, largely due to issues of reliability, quality, externalities, and the absolute number of interested parties. Although each country and city has their own site-specific situations, general observations are explained across low-, middle-, and high-income countries in The World Bank report 2012 (Table 1 of the Report).

1.3.4 Global policy trends

Global waste production has been on the rise due to different reasons in different parts of the world. Consequently the waste disposal has become a serious issue for several countries. Over the last twenty years, the countries with no proper policy structure have been formulating regulations and legal provisions to manage MSW. The main emphasis in almost all countries is on 'waste hierarchy', i.e., reduction of waste followed by reuse and recycle.

In North America, reduction and recycling dominate the management scenario, and in Australia, reduction, recycling and zero waste predominate. In Europe, the policy stress is on landfill diversion, and control of biodegradable waste to landfill. Some European countries control combustible waste ending up in landfill, and insist increased processing and reprocessing of waste. Japan, due to paucity of land space, insists on recycling and energy recovery at city level itself. The developing countries have recognised the urgent need to improve waste management practices as the rapid rise in their populations and economic activity create more and more waste.

After recycle and reuse, producing energy from waste is a good option when appreciable organic content is present.

Despite considerable progress globally in solid waste management practices in recent years, basic institutional, financial, social, and environmental problems still exist and need serious consideration.

1.3.5 Municipal waste, a source of clean energy

Municipal waste is abundantly available especially in consumer oriented societies, and is being viewed worldwide as a valuable commodity. It's considerable energy content and high fraction of biogenic carbon (50–60%) indicates high potential as a renewble energy source. Locally generated, waste can be utilized for the recovery of materials and energy rather than disposing in landfills creating environmental hazards (Figure 1.4). Waste can substitute or supplement fossil fuels in power generation and other industrial processes safeguarding energy supply.

With anticipated global shortages of critical nutrients such as phosphorus and increasing demand for renewable energy sources, the heating value and nutrient content of liquid and solid wastes are relevant for utilization (BC Ministry of Community Development 2009).

The Global Round Table on Climate Change, an initiative sponsored by The Earth Institute, Columbia University concluded that decarbonization of the economy could

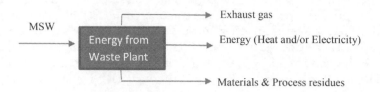

Figure 1.4 Energy and materials recovery from municipal solid waste.

be achieved through "the use of non-fossil-fuel-based sources . . . such as wind, geothermal, hydro, tidal, wave, nuclear, waste-to-energy, and/or biomass; and efforts to reduce global emissions of methane from landfills should be expanded, including increased use of waste-to-energy facilities where appropriate and cost-effective."

Energy-from-waste (EfW) meets the two basic criteria of a renewable energy resource, namely, its fuel source (i.e., municipal solid waste) is sustainable and indigenous. EfW facilities generate 'clean' energy from the garbage after the implementing the most preferred 'reduce and reuse' of waste. These facilities, therefore, deserve the same treatment as any other renewable energy resource (IWSA report 2007).

Energy from waste can be a part of a country's energy strategy/policy as it offers large potential for the mitigation of the negative effects of climate change.

1.3.6   Energy-from-Waste technologies

The essential property to decide whether the municipal waste can be used for energy recovery is its heating (calorific) value. It is a parameter strongly influenced by the composition of the waste which varies with the place and season. The study of macro-chemistry of municipal waste, i.e., the percentage of the major chemical constituents – carbon, hydrogen, oxygen, sulfur, nitrogen, chlorine, and ash – enables to estimate the heat content. This data is also required for estimating other critical parameters that decide the efficacy of energy recovery from waste.

The energy content of the waste is expressed as the lower calorific value (LCV) which varies widely from country to country. For residual MSW, LCV is about 2–5 MJ/kg in developing countries and 8–12 MJ/kg in industrialized countries. A LCV of 6 MJ/kg for the waste is needed for the safe operation of thermal treatment systems, and this level is reached in many countries

A good correlation exists between the LCV of the waste and the economic level of a country, namely GDP (IEA Bioenergy-Task 36, 2007–2009, End of Task Report).

To comply with process requirements, the 'raw waste' is also converted into Solid Recovered Fuel (SRF) or Refuse derived fuel (RDF) which have a lower contamination and better homogeneity. The preparation of RDF requires a basic level of treatment to remove recyclables from the waste stream, while SRF requires a higher standard of preparation. RDF is typically meant for standard EfW facilities which also accept unprepared mixed waste streams while SRF is typically used within cement kilns and power stations as an alternative to fossil fuels.

The chemical energy stored in the waste is extracted into diverse usable energy forms. This process is carried through two main pathways: thermal treatment and

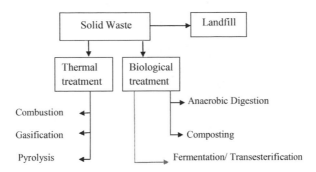

Figure 1.5 Waste treatment methods.

biochemical (biological) treatment shown in Figure 1.5. The waste feed-stocks include not only municipal solid waste, but construction and demolition debris, farm waste and livestock manure, industrial waste such as from coal mining, and the gases that naturally emanate from landfills.

The thermal and biological treatments of waste help basically to divert from land-fill large volumes of residual MSW which cannot be economically recycled through materials recovery or through treatment such as composting.

Combustion, gasification and pyrolysis are the three thermal processes utilized to treat the waste (e.g., Kolb and Seifert 2002, WSP 2013, Kokalj and Semac 2013). While combustion process was considered conventional, the gasification and pyrolysis were considered advanced thermal processes. Each process is characterized by the way the fuel is oxidized to produce energy.

When the municipal waste is introduced into a high temperature environment (a reactor), it first gets dried and then decomposes or pyrolyse into volatile (tars and gases) and char components. This stage is common to all three processes. How the process differs in each of them afterwards is as follows.

Combustion process uses an excess amount of oxidant (generally air) so the char and volatiles in the combustion reactor burn completely. The full calorific value of the fuel is released into the reactor and the sensible heat of the flue gases consisting primarily of carbon dioxide. This heat can be used to raise steam for power generation with a steam turbine. Combustion (also called Incineration) requires high temperatures for ignition, sufficient turbulence to mix all of the components with the oxidant, and time to complete all of the oxidation reactions.

Gasification process uses a limited supply of oxidant (oxygen), to maintain both combustion and reducing reactions in the same reactor. Some of the char and volatiles burn to supply the heat needed for pyrolysis and for further reactions that produce a synthesis gas (syngas) containing principally carbon monoxide and hydrogen, in addition to methane and other lighter hydro-carbons. The energy in the fuel is thus largely transferred into the heating value of the gas exiting the reactor, which is then burned in a gas turbine or engine to generate power or in a boiler to raise steam. Syngas can also be used as an intermediate for producing synthetic natural gas, methanol, dimethyl ether and other chemicals.

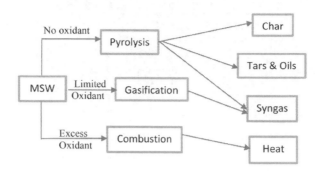

Figure 1.6 The three thermal conversion technologies and their end products.

Pyrolysis process utilizes no oxygen and the char and volatiles remain largely unchanged. The energy in the fuel (waste) is thus transferred to the heating value of the volatiles, and the char removed from the reactor. These can be burned separately in turbines, engines or boilers to generate power. The volatiles can be also condensed to give a liquid that can be used as a synthetic fuel. The proportions of gas, liquid and char produced depend upon the heating rate, the temperature of the reactor and so on. The heat for the pyrolysis reactor is usually supplied by burning some of the product gas in a separate heater. These are represented in Figure 1.6.

The end result in all the three is the same: while the energy in the waste is released by oxidation through combustion to produce heat that can be used to produce electricity, gasification and pyrolysis are a means of gaining more control over this process by converting the heterogeneous solid fuel into a consistent liquid or gaseous intermediate fuel (Howes 2012). Pyrolysis and gasification, considered as more advanced thermo-chemical conversion processes, have been in practice since 1970s (Kolb and Seifert 2002). These two processes and plasma-based technologies (discussed later) have been applied to 'selected' waste streams and on a 'smaller scale' than Combustion. The process conditions are strictly controlled in specially designed reactors in all the three processes, as shown in Table 1.3.

Each conversion technology sets different requirements for the input, employs different equipment configurations operating in different modes and gives a different range of products. Unlike in incineration, both pyrolysis and gasification may be used for recovering the chemical value of the waste, rather than its energetic value. The chemical products derived may be used as feedstock for other processes or as a secondary fuel as already mentioned. However, when applied to wastes, pyrolysis, gasification and combustion based processes are often combined, usually on the same site as part of an integrated process. In general, these types of integrated processes recover the total energy value rather than the chemical value of the waste, as in a conventional incinerator (Bosmans et al. 2013).

The waste thermal treatment processes are essential to a sustainable Integrated Municipal Solid Waste Management system (IMSWM) (Brunner et al. 2004; Porteous, 2005; Psomopoulos et al. 2009) as witnessed by the successful operating waste management systems. These processes are characterized by higher temperatures and conversion rates compared to others such as biochemical and physicochemical

Table 1.3 Characteristics of the main thermochemical conversion technologies (Kolb & Seifert 2002).

	Pyrolysis	Gasification	Combustion	Plasma
Objective	Maximize thermal decomposition of solid waste into coke, gases and condensed phases	Maximize waste conversion into high calorific fuel gases	Maximize waste conversion into high temperature flue gases	Maximize waste conversion into high calorific fuel gases and an inert solid slag phase
Temperature	250–900°C	500–1800°C	800–1450°C	1200–2000°C
Pressure	1 bar	1–45 bar	1 bar	1 bar
Atmosphere	Inert/nitrogen (no oxidant)	Gasification agent: O_2, H_2O	Air	Gasification agent: O_2, H_2O Plasma gas: O_2, N_2, Ar
Stoichiometric ratio	0	< 1	> 1	< 1
Products from the process:				
(i) Gas phase	H_2, CO, H_2O, N_2 and hydrocarbons	H_2, CO, CO_2, CH_4, H_2O, N_2	CO_2, H_2O, O_2, N_2	H_2, CO, CO_2, CH_4, H_2O, N_2
(ii) Solid phase	Ash, coke	Slag, ash	Ash, slag	Slag, ash
(iii) Liquid phase	Pyrolysis oil, water	–	–	–

processes. They allow an efficient treatment of diverse types of solid waste, mainly unsorted residual waste, and offer several advantages (Arena 2012):

(i) reduction of about 70–80% in mass and about 80–90% in volume, resulting in significant reduction of landfill space required (Consonni et al. 2005);

(ii) the land required for the EfW plant to process the waste is significantly smaller than that needed for landfill the same quantity of MSW (Psomopoulos et al. 2009);

(iii) organic contaminants such as halogenated hydrocarbons can be destructed (McKay 2002; Buekens and Cen 2011);

(iv) inorganic contaminants can be collected for useful and safe utilization or disposal (ISWA 2008; Samaras et al. 2010);

(v) utilization of recyclables such as ferrous and non-ferrous metals from bottom ash and slag (ISWA 2006; CEWEP 2011);

(vi) reduction of greenhouse gas emissions from anaerobic decomposition of the organic wastes (studies showed that about 1 ton of equivalent CO_2 is saved per each ton of waste combusted rather than landfilled, Psomopoulos et al. 2009);

(vii) better environmental performance and meet more severe emission regulations with respect to other energy sources (Arena et al. 2003; Azapagic et al. 2004); The power generation from EfW has been recognized to cause less environmental impact than almost any other source of electricity' (US EPA 2003; Rechberger and Schöller 2006);

(viii) 'clean' energy production, especially when the EfW plant is designed and operated for cogeneration of heat and power (CHP) (Rechberger and Schöller 2006; EC-IPPC 2006).

To sum up, thermal disposal methods particularly mass-burn incineration utilizing mature as well as innovative technologies, are a desired and viable option often used in industrialized countries to generate electricity and/or heat (Akehata 1998; Rylander 1997; Price 1996; Vehlow 1996; Hjelmar 1996; Bontoux 1999; Anon 2000; Faulstich and Jorgens 2001).

Almost 2240 waste-to-energy plants are active worldwide with a disposal capacity of 270 million TPA. It is estimated that almost 500 new plants with a capacity of about 150 million TPA to be installed by 2023 (Mark Doing 2015, ecoprog GmbH).

REFERENCES

Adam Vaughan (2013) News item. *The Guardian*, March 19, 2013.

Akehata, T. (1998) Energy recovery. *Macromolecular Symposia*, 135 (12), 359–373.

Annepu, R.K. (January 2012) *Sustainable Solid Waste Management in India*. Thesis submitted for MS in Earth Resources Engineering. New York, NY, Department of Earth and Environmental Engineering, Columbia University.

Anonymous (2000) *Profile Incineration in Europe*. Report prepared for Juniper for ASSURE.

Arena, U. (2012) Process and technological aspects of municipal solid waste gasification. A review. *Waste Management*, 32, 625–639.

Arena, U., Mastellone, M.L. & Perugini, F. (2003) The environmental performance of alternative solid waste management options. *Chemical Engineering Journal*, 96 (1–3), 207–222.

Azapagic, A., Perdan, S. & Clift, R. (2004) *Sustainable Development in Practice*. Chichester, UK, John Wiley & Sons Ltd.

Bontoux, L. (1999) *The Incineration of Waste in Europe: Issues and Perspectives*. Report, European Commission, Luxemburg, EUR18717EN.

Bosmans, A., Vanderreydt, I., Geysen, D. & Helsen, L. (2013) The crucial role of Waste-to-Energy technologies in enhanced landfill mining: A technology review. *Journal of Cleaner Production*, 5, 10–23.

Brunner, P.H., Morf, L. & Rechberger, H. (2004) Thermal waste treatment—A necessary element for sustainable waste management. In: Twardowska, A. & Kettrup, L. (eds.) *Solid Waste: Assessment, Monitoring, Remediation*. Amsterdam, The Netherlands, Elsevier B.V.

Buekans, A., Yan, M., Jiang, X., Li, X., Lu, S., Chi, Y., et al. (2012) Die thermische Abfallbehandlung in China (Thermal waste treatment in China).

Buekens, A. & Cen, K. (2011) Waste incineration, PVC, and dioxins. *Journal of Material Cycles and Waste Management*. doi:10.1007/s10163-011-0018-9.

CEWEP (2011) *Environmentally Sound Use of Bottom Ash*. Available from: www.cewep.eu.

CEWEP (2012) *Industry Barometer*. Available from: http://www.cewep.eu/media/www.cewep.eu/org/med_709/918_cewep_ecoprog_-_industry_barometer_wte_2012.pdf.

Cheng, H. & Hu, Y. (2010) MSW as a renewable source of energy: Current and future practices in China. *Bioresource Technology*, 101 (11), 3816–3824.

Consonni, S., Giugliano, M. & Grosso, M. (2005) Alternative strategies for energy recovery from municipal solid waste. Part A: Mass and energy balances. *Waste Management*, 25, 123–135.

Doing, M. (2015) Waste Industry, ecoprog. GmbH. Available from: www.ecoprog.com/en/publications/waste-industry.htm.

EIA. Available from: http://www.eia.gov/todayinenergy/detail.cfm?id=8010.

Energy Information Administration (2007) *Methodology for Allocating MSW to Biogenic/Non-Biogenic Energy*—Calculated from composition of waste and assigned energy values.

European Commission (2008) *Eurostat*. Available from: http://epp.eurostat.ec.europa.eu/portal/page/portal/waste/data/databaseg.

European Environment Agency State of the Environment Report 2012.

Eurostat (2008) *Generation and Treatment of Waste in Europe.*

Eurostat (2011) *Statistics in Focus 31/2011.*

Faulstich, M. & Jorgens, L. (2000) Incineration of residual waste in industrial processes in Germany. In: *Proc. 15th Intl. Conference on Solid Waste Technology and Management, 12–15 December, 1999, Philadelphia, PA.*

Hanjer. Available from: www.proparco.fr/jahia/webdav/.../SPD15_Irfan_furniturwala_uk.pdf.

Hanrahan, D., Srivastava, S. & Sita Ramakrishna, A. (May 2006) *Improving Management of MSW in India: Overview and Challenges.* New Delhi, Environment Unit, South Asia Region, The World Bank (India Country Office).

Hickman Jr., L. (2003) *American Alchemy: The History of Solid Waste Management in the United States.* Santa Barbara, CA, Forester Press.

Hjelmar, O. (1996) Waste management in Denmark. *Waste Management*, 16 (5/6), 389–394.

Howes, P. *IEA Bioenergy Task 36.* AEA Technology plc. Available from: www.ieabioenergy task36.org.

Howes, P. (2012). Municipal waste as a feedstock for next generation biofuels, Presented at FO Lichts Conference, Copenhagen, Feb. 2012.

Howes, P., McCubbin, I., Landy, M. & Glenn, E. (2007) *Biomass and Biofuels—A European Competitive and Innovative Edge.* Policy Department Economic and Scientific Policy. IP/A/ITRE/FWC/2006-087/Lot4/C1/SC1. 98 pp.

IEA Bioenergy. *SRF Seminar, Dublin, November 2011.* Available from: www.ieabioenergy task36.org.

IEA Bioenergy. *Measurement of Biogenic Content of Waste*, report in publication. Available from: www.ieabioenergytask36.org.

IEA Bioenergy. *Task 36: Final Report 2009.* Available from: www.ieabioenergytask36.org.

IEA Bioenergy. *Task 36: Integrating Energy Recovery into Solid Waste Management Systems (2007–2009).* End of Task Report, Chapter 4. Available from: www.ieabioenergytask36. org/..../2007-2009/Introduction_Final.pdf.

ISWA (2006) *Management of Bottom Ash from WTE Plants, ISWA-WG Thermal Treatment Subgroup Bottom Ash from WTE-Plants.* Available from: www.iswa.org.

ISWA (2008) *Management of APC residues from WTE Plants, ISWA-WG Thermal Treatment of Waste.* 2nd edition. Available from: www.iswa.org.

ISWA (2009) *Waste and Climate Change—ISWA White Paper.*

IWSA (2007) Report from: http://energyrecoverycouncil.org/userfiles/file/IWSA_2007_Directory.pdf.

Jayarama Reddy, P. (2011) *Municipal Solid Waste Management—Processes, Energy Recovery, Global Examples.* Hyderabad, CRC Press/Balkema, Leiden, & BS Publications.

Jayarama Reddy, P. (2014a) Municipal solid waste treatment in developing countries. In: *Encyclopedia of Environmental Management.* doi:10.1081/E-EEM-120051410, © Taylor & Francis, 2014.

Jayarama Reddy, P. (2014b) Municipal solid waste: Reuse and recycling. In: *Encyclopedia of Environmental Management.* doi:10.1081/E-EEM-120050626, © Taylor & Francis, 2014.

Jayarama Reddy, P. (2014c) Municipal solid waste: Energy recovery. In: *Encyclopedia of Environmental Management.* doi:10.1081/E-EEM-120050613, © Taylor & Francis, 2014.

Joseph, K. (2007) Lessons from MSW processing initiatives in India. In: *Proceedings of Intl. Symposium on MBT, 2007.*

Kleis, H. & Dalagar, S. (2007) *100 Years of Waste Incineration in Denmark—From Refuse Destruction Plants to High-Technology Energy Work.*

Kokalj, F. & Semac, N. (2013) Combustion of MSW for power production. In: Hoon Kiat Ng (ed.) *Advances in Internal Combustion Engines and Fuel Technologies.* InTechOpen. pp. 277–309. ISBN: 978-953-51-1048-4, March 20, 2003.

Kolb, T. & Seifert, H. (2002) Thermal waste treatment: State of the art—A summary. In: *Waste Management 2002: The Future of Waste Management in Europe*. Strasbourg, France (Düsseldorf, Germany), VDI GVC.

Lamers (2012) Emerging technologies for advanced thermal treatment of waste. In: *Next GenBioWaste 3rd International conference on biomass and waste combustion, London, 2012*.

Lecoste, E. & Chalmin, P. (2006) *From Waste to Resource—2006 World Waste Survey*. Economica Editions.

Malkow, T. (2004) Novel and innovative pyrolysis and gasification technologies for energy efficient and environmentally sound MSW disposal. *Waste Management*, 24, 53–79.

McKay, G. (2002) Dioxin characterization, formation and minimization during municipal solid waste incineration: A review. *Chemical Engineering Journal*, 86, 343–368.

Moses, K. (2013) News item. *The Guardian*, June 13, 2013.

Münster, M. (April 2009) *Energy Systems Analysis of Waste to Energy Technologies by use of EnergyPLAN*. Risø-R-1667(EN). Available from: http://130.226.56.153/rispubl/reports/ris-r-1667.pdf 3.

Murphy, J.D. & McKeogh, E. (2004) Analysis of energy production from municipal solid waste. *Renewable Energy*, 29, 1043–1057.

Niessen, W.R. (2002) *Combustion and Incineration Processes: Applications in Environmental Engineering*. 3rd edition. New York, NY, Marcel Dekker Inc.

Papageorgiou, A., Barton, J.R. & Karagiannidis, A. (2009) Assessment of the greenhouse effect impact of technologies used for energy recovery from municipal waste: A case for England. *Journal of Environmental Management*, 90, 2999–3012.

Planning Commission (May 2014) *Report of the Task Force on WTE (Volume I)*. New Delhi, Planning Commission of India.

Porteous, A. (2005) Why energy from waste incineration is an essential component of environmentally responsible waste management. *Waste Management*, 25, 451–459.

Price, B. (1996) *Energy from Waste*. London, UK, FT Energy.

Psomopoulos, C.S., Bourka, A. & Themelis, N.J. (2009) Waste-to-energy: A review of the status and benefits in USA. *Waste Management*, 29, 1718–1724.

Qui, L., Dong, Y. & Themelis, N.J. (2012) Rapid growth of WTE in China, Current performance and impediments to future growth. In: *Proc. 20th Annual North American Waste to Energy Conf., NAWTEC 20, April 2012*.

Rechberger, H. & Schöller, G. (May 18, 2006) Comparison of Relevant Air Emissions from Selected Combustion Technologies, Project CAST, CEWEP—Congress, Waste-to-Energy in European Policy.

Rylander, H. (1997) The Evolution of WtE utilization—The European perspectives. In: *Proc. 5th Annual North American Waste to Energy Conference, 22–27 April 1997, Research Triangle Park, N.C. 1-20, Conf. 970440*.

Sakai, S., Sawell, S.E., Chandler, A.J. & Eighmy, T.T. (1996) World trends in municipal solid waste management. *Waste Management*, 16 (5–6), 341–350.

Samaras, P., Karagiannidis, A., Kalogirou, E., Themelis, N. & Kontogianni, St. (2010) An inventory of characteristics and treatment processes for fly ash from waste-to-energy facilities for municipal solid wastes. In: *3rd Int. Symp. on Energy from Biomass and Waste, Venice, Italy, 8–11 November, 2010*. Italy, CISA Publisher. ISBN: 978-88-6265-008.

Stantec (2010) *Waste to Energy. A Technical Review of MSW Thermal Treatment Practices*. Final Report for Environmental Quality Branch Environmental Protection Division, Project No.: 1231-10166. Available from: www.env.gov.bc.ca/epd/mun-waste/reports/pdf/bcmoe-wte-emmissions-revmar2011.pdf.

Themelis, N.J. & Mussche, C. (July 2014) *2014 Energy and Economic Value of MSW, Including Non-Recycled Plastic, Currently Landfilled in the Fifty States*. Earth Engineering Centre, Columbia University.

Tolis, A., Rentizelas, A., Aravossis, K. & Tatsiopoulos, I. (2010) Electricity and combined heat and power from municipal solid waste; theoretically optimal investment decision time and emissions trading implications. *Waste Management and Research*, 28 (11), 985–995.

UN Statistics. Available from: http://unstats.un.org/unsd/environment/wastetreatment.htm.

UNEP (October 2009) RIM Background paper for consultations, 3rd Draft.

UNEP (2010) *Waste and Climate Change: Global Trends and Strategy Framework*. Osaka/Shiga, Japan, UNEP, Division of Technology, Industry and Economics, IETC.

U.S. EPA (2003) Letter to President of Integrated Waste Service Association. Available from: www.wte.org/docs/epaletter.pdf.

U.S. EPA (2009) *Municipal Solid Waste Generation, Recycling, and Disposal in the United States: Facts and Figures for 2009*. Technical Report, 2009.

U.S. EPA (2011) Waste Management Hierarchy.

U.S. EPA (May 2013) *Municipal Solid Waste in the US: 2011 Facts & Figures*.

U.S. EPA (February 2014) *Municipal Solid Waste Generation, Recycling, and Disposal in the United States: Facts and Figures for 2012*. EPA-530-F-14-001

U.S. EPA (June 2015) *Advancing Sustainable Materials Management: 2013 Fact Sheet. Assessing Trends in Materials Generation, Recycling and Disposal in the U.S.*

Vehlow, J. (1996) MSW management in Germany. *Waste Management*, Vol. 16, Issues 5–6, pp. 367–374.

Waste Management World. *Waste to Energy in Brazil*. Available from: http://www.waste-management-world.com/index/display/article-display/1476541567/articles/waste-management-world/volume-12/issue-3/features/wte-the-redeemer-of-brazils-waste-legacy.html.

Waste Management World (July 30, 2012) Global Municipal Solid Waste to double by 2025.

Williams, P. (2005) *Waste Treatment and Disposal*. 2nd edition. Chichester, UK, John Wiley & Sons Ltd.

Wilson, B., Williams, N., Liss, B. & Wilson, B. (October 2013) *A Comparative Assessment of Commercial Technologies for Conversion of Solid Waste to Energy*. EnviroPower Renewable, Inc.

Wilts, H. (2012) National waste prevention programs—Indicators on progress and barriers. *Waste Management Research*, 30, 29–35.

World Bank. *Waste Management in South Asia*. Available from: http://ppp.worldbank.org/public-private-partnership/sitesppp.worldbank.org/files/documents/ADB_sustainable-waste-management-south-asia.pdf.

World Bank (1999) *Municipal Solid Waste Incineration*. World Bank Technical Guidance Report. Washington, DC, World Bank.

World Bank (May 2005) *Waste Management in China: Issues and Recommendations*. Urban Development Working papers, East Asia Infrastructure Department, Working paper 9.

World Bank Report (March 2012) *What a Waste: A Global Review of Solid Waste Management*. Hoornweg, D. & Bada-Tata, P., World Bank Report, No. 15.

WSP (2013) *Review of State-of-the-Art Waste-to-Energy Technologies*. Stage 2—Case Studies prepared by Kevin Whiting. London, UK, WSP Environmental Ltd, WSP House.

Yadav, D. (2009) Garbage disposal plant mired in controversies. *India Times*, February 19, 2009.

Zerbock, O. (2003) *Urban Solid Waste Management: Waste Reduction in Developing Nations*. MS Thesis. Michigan Technological University.

Zhang, D.Q., Tan, S.K. & Gersberg, R.M. (2010) Municipal solid waste management in China: Status, problems and challenges. *Journal of Environmental Management*, 91, 1623–1633.

Zhou, J. & Chen, H. (2012) MSW incineration in China: the current practices and future challenges. In: *2012 Intl. Conf. on Future Electrical Power and Energy Systems*, Lecture Notes in Information Technology, Vol. 9.

Chapter 2

Combustion technology

2.1 INTRODUCTION

Combustion technology is perhaps the oldest for recovering the energy stored in MSW. The most prevailing incineration technology is the '*Moving Grate*', which is designed to handle large volumes of MSW with essentially no pre-treatment. The grate incinerators are often referred to as 'mass burn incinerators (MBI)'. Moving grate type engages large-scale combustion in a single-stage chamber unit where complete oxidation occurs. In the mass burn incinerators, the thermal energy generates electricity through steam turbines. In incinerators where both heat and electricity are generated, the recovered residual heat is used for district heating, hot water supply, and so on (Williams 2005, EC 2006, Defra UK 2013). The other combustion-based technologies are '*Fluidized bed*' combustors. Both moving grate and fluidized bed are well established processes with several thousands of plants successfully operating globally.

Europe has a lot of experience with Waste incineration, and Denmark and Sweden have been leaders in using the energy generated from incineration for more than a century. This is due to land resource issues and higher overall thermal efficiencies where heat rejected in the power cycle can be used and not just transferred to the environment (atmosphere or water). In 2005, waste incineration produced 4.8% of the electricity consumption and 13.7% of the total domestic heat consumption in Denmark (Kleis and Dalagar 2007). Switzerland and Norway, along with Sweden, which rank high in the Environmental Performance Index (2008), combust almost all of their non-recyclable wastes. These countries have rigorous recycling policies, which play a vital role in the environmental impacts of waste combustion. A number of other European countries including Luxembourg, the Netherlands, Germany and France, rely heavily on incineration for handling municipal waste, in particular (EC 2006, EnviroPower 2013). The majority of new WTE facilities are based on mass burn systems. In 2001, 450 conventional combustion facilities (420 mass burn, and 30 fluidized bed) were operating in Europe (Stantec Final report 2011). In mid-2013, approximately 520 WtE plants were operational in Europe. They were able to treat around 95 million tonnes of MSW and commercial waste annually. Over the past five years, the European WtE capacity grew by an annual treatment rate of 19 million tonnes (24%). In the same period, 73 new WtE plants were commissioned, while only eight older facilities were shut down. All in all, Europe saw a steady and strong growth of WtE treatment capacities for more than 10 years (Mark Doing 2013). Recent projections developed by the European Confederation of Waste to Energy Plants (CEWEP) show that over 1020 plants (with a combined capacity of 177 million tons/year) will be in operation in Europe by 2016.

According to USEPA, in 2010, 86 waste-to-energy facilities were in operation in the U.S. (63 massburn operating at 25% efficiency, 16 RFD, and 7 modular). The report further says: massburn/waterwall, massburn/modular, and RDF/dedicated boiler technologies are 'commercially proven' and their risk levels are 'very low risk', 'low' and 'low' respectively. RDF/fluidized Bed is a 'proven' technology with 'moderate' risk. Gasification has 'limited operating experience' with 'high' risk, whereas pyrolysis has 'uncertain commercial potential' with 'high' risk (Miller of U.S. EPA 2011).

In Asia, there is limited experience with waste incineration for energy recovery outside the industrialized countries of Japan, Singapore, and Taiwan (World Bank 1999). In India, eight Waste to Energy plants are set up in 2012, and quite a few of them are non-operational (Planning Commission TF report 2014). Two refuse derived fuel plants (capacity 50–700 t/d) started generating power. But they were inconsistent in operation and presently not working (Annepu 2012, Jayarama Reddy 2011, Yadav 2009, Joseph 2007). In China, 74 waste-to-energy plants were working in early 2000 treating 40,020 t/d; and many more plants were in planning in 2008 (Zhang et al. 2010, Qiu et al. 2012, see also Liu Y. 2005, Cheng et al. 2007).

2.2   BENEFITS & ISSUES

Combustion of municipal solid waste evolved from a means of reducing the volume of waste through uncontrolled burning to a sophisticated technology with more objectives. It is an established procedure for disposal of MSW and can be constructed closer to the MSW sources or collection points reducing transport costs. By heat recovery, the cost of the operation can be compensated by energy sales (Brunner 1994).

Nevertheless, the main problems associated are the large volume of gaseous emissions containing heavy metals including mercury and highly toxic dioxins and furans which pose environmental health risks (Moy et al. 2008) and hazardous solid wastes that remain after incineration as fly ash or air pollution control (APC) residues (Quina et al. 2008a,b, Vehlow 2015). However, knowledge of the microchemistry of waste, namely, the content of heavy metals and other environmentally vital species present at ppm level enables to assess the potential pollutant emission problems (Niessen 2002).

Over the last 10 to 15 years, the incineration sector has undergone rapid technological development. Much of this change has been driven by legislation in EU, the U.S. and so on, specific to the industry. The application and enforcement of modern emission standards stands as an example. The use of modern pollution control technologies has reduced emissions to air to levels at which pollution risks from waste incinerators are now considered to be very low, although this has led to high capital costs (BREF 2006). With continuous process development and process control, application of Computational Fluid Dynamics (CFD) approach in the plant design, and the development of resource recovery options from the solid residues, the sector has been developing techniques which limit operating costs and at the same time maintaining or improving environmental performance. The 4th generation WtE plant built by Amsterdam's Afval Energie Bedrijf (AEB) in the Netherlands offers a foremost example of how incinerators can attain both high energy and materials recovery. The net electrical efficiency is 34.5% as agaist the 22 to 26% net electrical efficiency of current state-of-the-art incineration plants. Process data obtained for one year of operation (09/2009-09/2010) show a net electrical efficiency between 20% and 31% (Van Berlo 2010) which confirms it is possible to meet the high expectations (Bosmans et al. 2013). The other newly built incinerators for electricity production showing high efficiencies are: the MKW Bremen plant, 30.5%; EVI Laar, 30.5%; AZM Moerdijk, 32.5%. In the case of combined heat and power production (CHP), the net electrical efficiency is close to 23% and its thermal efficiency is around 45%, which is technically possible by using the back pressure turbine technology (Tolis et al. 2010). Some of these Plants are briefly described under 'Case Studies'.

In addition to the recovery of energy either as heat or electricity or combined heat and power (CHP), the technology allows recovery of metals that would otherwise be lost from the productive lifecycle of material goods.

2.3 CHEMISTRY OF COMBUSTION

Municipal solid waste, unlike many fossil fuels, is more heterogeneous and the physics and chemistry of its combustion are very complex. The combustion process involves chemical reaction kinetics and equilibrium, combustor fluid mechanics, and heat transfer rates. Hence a large number of reactions of complex nature are implied in its complete oxidation at high temperatures. For this reason, the reaction mechanisms are not yet fully understood (Jansson 2008). Combustion systems are, therefore, multipart involving simultaneously linked heat and mass transfer, chemical reactions and fluid flows.

The two significant combustible chemical elements in the waste are carbon and hydrogen. Chlorine and sulfur are usually of minor significance as sources of heat, but are the major constituents, mainly chlorine, concerning corrosion and pollution.

When carbon and hydrogen are completely burned with oxygen, the following happen:

$$C + O_2 = CO_2 + \text{Heat released} \tag{2.1}$$

$$2H_2 + O_2 = 2H_2O + \text{Heat released} \tag{2.2}$$

Here, the stoichiometric ratio (SR) defined as the ratio of the actual amount of oxygen supplied in the oxidation process to the amount actually required is 1. The heat released increases the temperature of the respective products of combustion and the other gases present. The burning of compounds containing oxygen require less air since the compound already contains some oxygen that will be made available during the combustion process.

Jenkins et al. (1998) has given an empirical formula that represents the combustion of actual *wastes* in air. This chemical representation can be relevant to the municipal solid waste at any location:

$$\begin{aligned}
&C_{x1}H_{x2}O_{x3}N_{x4}S_{x5}Cl_{x6}Si_{x7}K_{x8}Ca_{x9}Mg_{x10}Na_{x11}P_{x12}Fe_{x13}Al_{x14}Ti_{x15} + n_1H_2O \\
&\quad + n_2(1+e)(O_2 + 3.76N_2) \rightarrow n_3CO_2 + n_4H_2O + n_5O_2 + n_6N_2 + n_7CO \\
&\quad + n_8CH_4 + n_9NO + n_{10}NO_2 + n_{11}SO_2 + n_{12}HCl + n_{13}KCl \\
&\quad + n_{14}K_2SO_4 + n_{15}C + \cdots
\end{aligned} \tag{2.3}$$

This empirical formula includes only 15 elements whereas the actual municipal solid waste may contain many more, some of them found in traces; and the molar indices x1 to x15 can vary widely. n_1 corresponds to the moisture in the waste; n_2 is related with the amount of air (considered as a binary mixture of O_2 and N_2) used in the combustion; $(1 + e)$ is the excess air in relation to the stoichiometric amount, usually ranges from 1.2 to 2.5 (depending on whether the fuel is gas, liquid or solid) (BREF 2006); n_3 to n_{15} correspond to the stoichiometric coefficients of the different species that can be found as reaction products, among many others that can be released in the emissions.

If the incinerated material is represented by a simpler formula, like $C_u H_v O_w N_x S_y$, then the combustion equation may be simplified and represented as

$$C_u H_v O_w N_x S_y + [u + (v/4) - (w/2) + y]O_2$$
$$\rightarrow uCO_2 + (v/2)H_2O + (x/2)N_2 + ySO_2 \tag{2.4}$$

Since air is the usual source of the oxygen, excess amounts of air will dilute the gases and reduce the temperature of the gases. The 'processed' MSW (i.e., after removing non-combustibles and high-moisture constituents) improves the fuel quality thereby increasing the heating value of the remaining waste and the specific air demand. In incineration, the energy is released through the oxidation reactions, and its recovery occurs directly from the gases (called flue gases) formed (Quina et al. 2011).

2.4   EFFICIENCY OF COMBUSTION

The prime objectives of combustion in an incinerator are (i) complete destruction of the organic constituents in the waste to form harmless (flue) gases and preventing the release of harmful material to the environment, and (ii) efficient conversion of the heat released by the flue gases into useful energy. The oxidation of the combustible waste primarily requires (i) a high *temperature* sufficient to ignite the constituents, (ii) mixing of the material with oxygen (which is called *turbulence*) and (iii) adequate *time* for the combustion to complete. These "three Ts" influence the combustion efficiency. If these three factors are properly addressed, the conversion efficiencies of 99.90%–99.95% can be achieved in well-operated incinerators (e.g., Guyer). High-efficiency oxidation of a combustible material requires the presence of more than the stoichiometric volume of oxygen, and it is one of the several factors that influence the rate of oxidation.

Generally, combustible gases and vapors require less excess oxygen to achieve high-efficiency oxidation due to the ease of mixing and the nature of the constituents in the gases and vapors. In contrast, solid fuel materials require more excess air and time because of the more complex processes involved in their combustion. Increasing the excess air quantity *beyond* the required amount does not promote the combustion process, but lowers the gas temperature thereby reducing the efficiency of the downstream heat-recovery process. The best combustion efficiency and energy recovery in a large water-wall type solid waste incinerator was observed with a system SR of 1.4 to 1.5. The secondary combustion chambers in modular and packaged incinerators achieve their highest oxidation efficiency at SR of 1.5 to 2.0 (Guyer).

2.5   PROCESS STABILIZATION & COMBUSTION CONTROL

For over 50 years, the focus has been on utilizing the hot flue gas from combustion of MSW to generate process steam and/or electricity. This has necessitated stabilizing the combustion process and operating the turbine most efficiently to maximize energy production. The *stabilized* combustion also contributes to lowering pollutant emissions. Therefore, from the point of both economics and environment, it is significant to stabilize the combustion process in the MSW incineration plant. This is done using an advanced control and monitoring system (CMS), a vital component in the incineration plant.

The wide variability in physical and chemical properties of MSW feedstock is the principal reason for the 'unstable combustion' in the chamber (Kilgroe et al. 1990), a major problem encountered in a mass-burn incineration facility. Other reasons for incomplete combustion noticed are (a) the transient phenomena such as the break-up of waste particles and the channelling of combustion air/oxygen, and (b) poor mixing of the flue gas or quenching of the combustion process by a heat sink such as a solid surface. Such a situation (unstable combustion) creates operating problems such as (i) instable chamber temperature, (ii) low thermal efficiency of the combustion chamber, and (iii) generation of soot, non-combustible residues and products of incomplete combustion (PICs) affecting the combustion efficiency significantly.

In order to maximize waste throughput and thereby the energy output while still fulfilling lifetime related and environmental conditions, the combustion *controllers* are used. This leads to a 'constraint-pushing' type of control behaviour where the aim is to operate as closely as possible to the dominating constraint (either environment or lifetime related) while violating this constraint a minimal number of times. The retreat to this constraint is determined by the capability of the combustion control system to suppress the 'variations' that are present on the controlled variables. These variations are very large for MSW combustion plants due to considerable inconsistency in the waste composition. Reducing the size of these variations reduces the retreat and allows for an economically more profitable operation. Particularly for the plants processing large amounts of waste, each small gain in 'retreat' allows a large increase in profit.

Conventionally, a *multivariable PID control system* (Thome-Kozmeinsky 1994) is used in the waste combustion plant to control the air flow rate, waste-feed rate and grate speed which affect the oxygen content in the flue gas as well as steam production. The 'Proportional-integral-derivative (PID)' controller is the most common control algorithm used in industry and consists of three basic coefficients: proportional, integral and derivative which are varied to get optimal response. None the less, this system is considered to have limited capability in exploiting the interactions present in the MSW combustion process. There were several performance-related studies on PID-controllers. Baxter and El Asri (2004) have optimized the existing PID control strategy as well as the parameters in the PID-controller, and shown that it is possible to make PID control strategy adapt to the actual working conditions by the use of gain scaling the set points. Studies by Thomsen and Lundtorp (2008) at the state-of-the-art FASAN MSW Incinerator at Næstved in Denmark, which includes analysis of the data with time series analysis, show that optimizing the PID control system can improve the stability of the steam production in many ways. The new PID structure not only minimizes the variations in the steam production using the oxygen controller

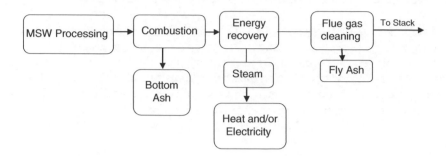

Figure 2.1 Flow diagram of Conventional MSW Incinerator.

more actively but is more robust in handling low calorific value without increasing the risk of a 'load breakdown'. The analysis of the impulse response shows that dynamics have changed to a response which tends to be more like a first order response, which makes it easier for a PID controller to cope with. An account of advanced automatic combustion control systems utilized in solid waste incinerators is given in Annexure 1.

2.6 MSW INCINERATOR SYSTEMS

A flow diagram of a conventional MSW Incineration system is shown in Figure 2.1. The conventional technologies are mainly *mass burn* incineration and *fluidized bed* incineration. Mass burn incineration compared to fluidized bed is the most dominant technology and commonly operated.

A mass burn system uses a hearth or *a Grate* to support a large mass of raw or processed waste as it is progressively burned down. The burn down process typically requires a nominal four to six hours from the time the waste is introduced into the primary combustion chamber until the ash is discharged. Incinerators operating under oxygen-deficient conditions require longer burn-down times than the furnaces operating in the oxygen-rich condition.

A schematic diagram of a typical municipal waste incinerator is given in Figure 2.2a. The incineration plant consists of several sub-systems (Quina et al. 2011): (a) waste reception and handling (storage, on site pre-treatment facilities), (b) combustion chamber (waste and air-supply systems), (c) energy recovery (boiler, economizer, etc.), (d) facilities for clean-up gaseous emissions, (e) on-site facilities for treatment or storage of residues and waste water, and stack, (f) devices and systems for controlling incineration operations, and for recording and monitoring incineration conditions.

The incinerator designs reflect a few aspects of the technologies earlier developed to burn wood, peat and lignite. However, application of spreader stoker and suspension burning techniques and some aspects of furnace design and fuel feeding are not yet included into conventional incinerator practice. Unlike the fuels mentioned, the municipal solid waste presents some unique problems to the furnace designer, mainly due to its high ash and moisture content, and relatively low energy density (Niessen 2002).

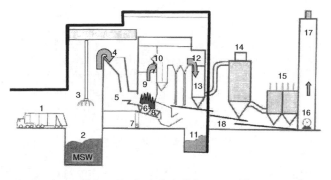

1 – waste collection vehicle
2 – waste storage pit
3 – waste handle crane
4 – feed hopper
5 – feeder
6 – grate

7 – forced–draft fan
8 – undergrate air zone
9 – furnace
10 – boiler
11 – bottom ash bunker
12 – superheater

13 – economiser
14 – fdry scrubber
15 – fabric filter baghouse
16 – induced–draft fan
17 – stack
18 – APC residues conveyor

Figure 2.2a Simplified scheme of a MSW incinerator (Quina et al. 2011, INTECH).

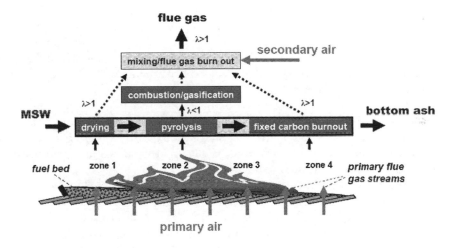

Figure 2.2b Idealized scheme of MSW combustion in a Grate furnace (Source: Hunsinger 2010; reproduced with the permission of the author, Hans Hunsinger, ITC, KIT, Germany).

2.6.1 Grate technology

Plant scheme & process

Grate incinerators are widely applied for the incineration of *mixed* municipal wastes, particularly low calorific (LCV) waste. The scheme of MSW combustion in a Grate furnace in an idealized scheme is shown in Figure 2.2b (Hunsinger 2010). Waste is collected and transported to the bunker for storage. The bunker functions as a waste buffer as well as a place to detect dangerous or incompatible materials that should be removed (Bilitewski et al. 1997). With the help of a crank, the waste is mixed, sorted and fed to the incinerator. The frequency and time of feeding the waste to the incinerator are very critical as they directly influence the incinerator's efficiency.

Due to its location between the bunker and the furnace, the feeding unit has to be robust enough to withstand the mechanical and thermal stresses.

The waste is slowly and continuously propelled through the combustion chamber by a mechanically actuated grate. The waste on the grate moves through different temperature zones as it procedes, and its thermal destruction occurs through sub-processes: drying, volatilization (pyrolysis and gasification), combustion and burnout of the residue. These processes may be superimposed on each other because of mostly insufficient mixing of the waste bed moving on the grates (e.g. Kreith 1959, Chapman 1986, Bardi and Astolfi 2010, Hunsinger 2010, Kokalj and Semac 2013, Fleck 2012). It is preferable to have the waste spread out evenly to maximize combustion. Hence, the grate systems are designed to have speed adjustments. The grate is sped up when highly combustible waste is fed, and slowed down when waste is wet or lower heat-value.

Drying – volatile content is evolved (hydrocarbons and water) at temperatures generally between 100 and 300°C. The drying zone, the front area of the grate, is important of the process since it influences the location of the combustion area on the grate. The onset timing of the combustion, which depends on the temperature distribution in the furnace, residence time of the flue gases, and the level of the material stress of the combustion bed grate, if too soon or too late, then the temperature distribution in the furnace may not be homogeneous, affecting the drying process which is strictly related to the heat transfer in the waste bed. As the waste bed on the grates is like a quasi-homogeneous bulk streamed by a fluid, the heat transfer (and hence the drying) is a function of the fluid velocity and its heat capacity, density, heat conductivity and viscosity (Kreith 1959, Chapman 1986, Bardi and Astolfi 2010).

Pyrolysis and gasification (Volatilization) – As the waste gets dried, the temperature of the waste increases, and at a particular value, the combustion under deficit oxygen ($\lambda < 1$, i.e., gasification) starts. In a shortwhile pyrolysis starts occurring under no oxygen when the higher hydrocarbons, $C_xH_yO_z$, are cracked into gaseous species such as CH_4, CO and H_2 at high temperatures.

Therefore, the primary airflow through the waste bed is kept below stoichiometric levels. This is the area of the grate with greatest carbon conversion. The energy for the pyrolysis is supplied by the high temperature reactions between oxygen and some of the volatile hydrocarbons and solid components. Now, a mixture of gasified hydrocarbons, water from the oxygen-hydrogen reaction and the remaining nitrogen leave the waste bed and undergoes homogeneous secondary combustion above the waste bed (Bardi and Astolfi 2010). The final composition of the pyrolysis gases and the volatilization ratio depend on the composition and temperature of the waste, and the residence time of the hot gas in the waste bed. After *volatilization,* the residual carbon called char starts combusting through a series of heterogeneous reactions followed by ash *burnout* at high temperatures.

The *complete process* can best be explained as follows: the fuel (waste) forms a bed on top of the grate, and the primary combustion air is injected through the multiple inlets or 'plenums' in the grate by a forced-draft fan. The primary air supply must be adequate to cool the grate, and to sustain the combustion. Normally, the former quantity is the larger. By partly cooling the grate by water (water-cooled grate), it is possible to adjust the primary air supply precisely to the flow needed for the primary combustion process only. The primary air is generally taken from the storage pit to bring down the air pressure and eliminate odour emissions from the storage area.

At this point, the waste is burned in sub-stoichiometric conditions, where the available oxygen is about 30 to 80% of the amount required for complete combustion, which results in the formation of pyrolysis gases. These gases are combined with excess air (secondary air) introduced into the combustion chamber through rows of high pressure-nozzles located in the side-walls of the chamber above the waste bed. The secondary air flow facilitates complete combustion of the flue gases by introducing turbulence for better mixing and by ensuring surplus of oxygen. Optimizing the direction of the secondary air jets also assists turbulence. The secondary air can be fresh air, or alternately, the flue gas can be mixed with air and recirculated. Approximately 10–20% by volume of the usually cleaned flue gases replaces the secondary air and the mixture recirclated (FGR). This technique is reported to reduce heat losses with the flue-gas, and to increase the process energy efficiency by around 0.75–2% (European Commission, 2006). Also it reduces the nitrogen oxide content in the flue gas by another 20% (Bilitewski et al. 1997).

It is also important whether the entire gas flow through the furnace is counter flow, centre flow or parallel flow as explained in the later pages. Thus, the air supply and the volumetric mixing of primary flue gases with secondary air is a delicate design feature.

It is pertinent to note that air injected below the grate (underfire air) enhances drying and primary combustion of the waste; and air injected above the grate (overfire air) assists in secondary combustion of gaseous volatilized organic compounds and 'products of incomplete combustion (PICs)' released by the waste burning on the grate.

Air and waste delivery rates synergistically determine combustion chamber temperatures. A properly operating facility maintains combustion temperatures between 800° and 1100°C. When excess air is introduced into the combustion chamber, temperatures fall resulting in carbon monoxide emissions. Excessive waste flow also decreases temperature, and results in incomplete combustion of the waste. Therefore, continuous emissions monitoring facility is used to control/regulate waste delivery rates and air flow to the combustion chamber to optimize the performance. The firing power control can be supported through additional sensors such as IR camera, pyro-detectors, and heating value sensors (e.g., Beckmann and Scholz 2002, Manca and Rovaglio 2002, Keller et al. 2007).

In more advanced systems, primary air is injected into the drying, burning and burnout zones separately from secondary air. The control of air injection is done either via manual adjustments to dampers or automatic combustion controls. The latter is a more advanced system that senses, via continuous monitoring devices, the temperature and oxygen needs of the furnace, and accordingly adjusts the quantity of primary and secondary air entering the furnace. In most of these advanced systems each of the three primary air zones is subdivided into a number of sections, each of which can be individually controlled. This type of system is designed to optimize the placement, velocity, and flow of air to all parts of the grate area, and improve combustion and reduce particulate generation. Babcock Wilcox Volund is one of the commercial developers of such an advanced combustion control system.

Modern plants are also configured to achieve improved combustion efficiency by utilizing highly engineered designs which include arches and bull noses. Arches are located above the burning and burnout zones to contain and funnel the combustion gases into the furnace proper. Bull noses are built into the furnace walls, usually near the point of injection of overfire air (secondary air), several feet above the grate for the purpose of introducing turbulence and retarding movement of the combustion gases out of the furnace. These features were designed for guiding the combustion gases through a more circuitous path which also, not incidentally, allows the gases greater prospect to mix with oxygen and permits them a longer 'residence time' in the high temperature region of the furnace (Clarke 2002).

In fact, 'in-furnace' steps such as air distribution and furnace design can influence those stages in order to reduce pollutants in gaseous emissions (BREF 2006). About pollutants, it must be noted that combustion includes very fast reactions occurring in gas phase (fraction of seconds), and self-supporting combustion is possible if heat value of the waste and oxygen content are adequate. Therefore, the grate length should ensure the completion of the phases. The speed with which the phases change from one to the other is determined by the waste composition and its heat value.

The same processes take place in all single-stage combustion facilities, although the local distribution and the time scales differ.

If the waste is completely incinerated, the resulting flue gas would contain water vapor, N_2, CO, and O_2. The reactions being exothermic, release a high amount of energy that is carried over by the flue gas as heat; and the recovery of heat occurs in the boiler and in the convective sections (superheater and economizer) of the incineration plant.

Since the waste stream is so heterogeneous, depending on the combustion conditions, other compounds are also formed or remain, and lifted upward off the grate by the heat of combustion. These include unburned carbon particles, partly burned compounds such as CO, polycyclic aromatic hydrocarbons (PAHs), PCBs, the more toxic dioxins and furans, incombustible elements such as heavy metals, and compounds such as HCl, HF, HBr, SO_2, NO_x and oxidized metals. Depending on the combustion temperatures during the main phases of incineration, volatile heavy metals and inorganic compounds (e.g. salts) are totally or partly evaporated, and are transferred from the input waste to both the flue gas and the fly ash it contains. A mineral residue fly ash (dust) and heavier solid ash (bottom ash) are created.

The residues from Waste incinerator can be grouped as follows: (a) Bottom ash as discharged from the bottom of the furnace (mainly the grate) and fallen through the furnace grates; (b) Heat recovery ash (HRA) as collected in the heat recovery system including boiler, economizer and superheater. HRA is frequently discharged into the bottom ash stream and thus is often included in a broader definition of bottom ash; (c) Fly ash carried over from the furnace and removed before cleaning the fuel gases; (d) Air pollution control (APC) residues as collected in the APC equipment (i.e. scrubbers, electrostatic precipitators, and baghouses) including fly ash, sorbents, condensates and reaction products. The term "fly ash" usually includes APC residues; (e) Combined ash as a mixture of the above categories. The amount of each of these residues depends on several factors: feed waste composition, incinerator technology and operation, and air pollution control system technology and operation (Kalogirou et al.)

As the air pollutants are harmful for the environment, stringent regulatory limits are set for their release into the atmosphere. Thus, depending on the desired emission limits, air pollution control (APC) system with different types of cleaning units is included in the Plant. A dry scrubber and fabric filter only are shown in Figure 2.2a. The flue gas cleaned off the pollutants is finally released via the stack by using an induced draft-fan.

Typically an input of 1 ton of MSW originates nearly 300 kg of bottom ashes, 30 kg of APC residues, and the rest is emitted as flue gas (Quina et al. 2011). Although the air flow rate is controlled based on the characteristics of the stack gases, generally, about 4000 to 4500 m^3/t of air is required to assure a fully oxidizing atmosphere (IAWG 1997). The flue gas volume depends on technology, particularly the provision of flue gas recirculation. The volumes may be in the range of 4500 to 6000 Nm^3/ton of waste though large local variations can be observed (Achternbosch and Richers 2002, BREF 2006).

For an in-depth account of the theory and practice of municipal solid waste combustion, the reader may refer to published papers/reviews/books; for example, Walter Niessen (2002).

Grate designs

Grate performance is critical to the entire plant performance; therefore the grate type, design and dimensioning should be carefully selected. The design should be suitable to accept a wide variation and possible changes in the calorific value and composition of the input waste. Furthermore, regardless of the specific properties and varying 'quality' of the waste, the grate should be able to meet all requirements for waste capacity, operational reliability, combustion efficiency, and operation at partial load.

Several designs of *grate firing systems* (furnaces) are available, though the first two given here are the mostly utilized designs.

(a)   Roller grate – consists of a series of drums or rollers in a stepped formation, with the drums rotating in the direction of the waste movement (Fig. 2.3)

(b)   Reciprocating grate – resembles stairs with alternating fixed or moving grates; pushes the waste in the direction of waste flow or against the flow in an upward direction (Fig. 2.3)

(c)   Rocking grate – pivoted or rocked grate producing an upward or forward motion, pushing the waste down the grate

(d)   Circular grate – a rotating annular hearth or cone agitates the waste

(e)   Rotary Kiln – an inclined cylinder as it rotates, causes a tumbling action to expose unburned material and advance the waste down the length of the kiln.

An important aspect of the incineration grate design is to ensure a proper and sufficient distribution of the incineration air into the furnace for complete combustion to occur. Generally, a primary air blower introduces air through the holes in the burning grates into the fuel bed. The primary combustion air also cools the grate itself (if it's air-cooled grate) as cooling is important for reducing thermal stress of the grate. The influence of the waste composition and parameters such as primary air, temperature etc. on combustion on the grates of different geometry and the related technological issues are very vital and are discussed elsewhere (e.g., Beckmann and Scholz 2002).

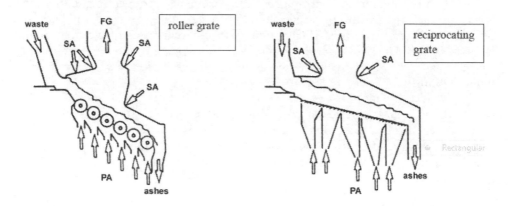

Figure 2.3 Grate types for mass-burn incinerator systems; PA: primary air, SA: secondary air; FG: flue gas (Source: Goerner 2001; Reproduced with the permission of Prof. Dr. Klaus Goerner).

Several types of commercial grates along with their engineering details developed in Europe and U.S. are covered in many publications (e.g. Walter Niessen 2002, Lannier Hickman Jr. 2003, Grate manufacturers' brochures).

Furnace design

Subsequent to the combustion of the waste on the grate, the resultant gases of combustion are completely burnt in the upper combustion chamber. It is therefore essential that the grate and furnace are designed as a fully integrated unit and not regarded as separate supply items.

Depending on grate design, the furnace geometry and the method of injecting secondary air are very vital for optimizing combustion.

The requirements that determine the design are: form and size of the incineration grate, vortexing and homogeneity of waste gas flow, sufficient residence time for the waste gasses in the hot furnace, and partial cooling of waste (BREF 2001).

In order to avoid fusion of the flying ash at the boiler, the waste gas temperature must not exceed an upper limit at the incineration chamber's exit.

The grates are made of special alloys to withstand the high temperature and to avoid corrosion. Globally, high quality moving grates, the most expensive part of the Plant equipment, are manufactured and supplied by Martin GmbH, Babcock & Walcox Volund.

In general, three standard furnace designs for MSW incineration plants have been established: parallel flow, counter flow, and centre flow (Fig. 2.4). In parallel flow design, the gas flow is in the same direction as the waste flow; in counter flow the gas flow is in the opposite direction of waste flow; and centre flow as shown in the Figure 2.4. Since the flue gas flows through hottest zone in the parallel flow design, the burn-out of the gaseous species is well done and hence is suitable for lower LCV waste. For the parallel- and counter-flow designs, heat transfer from the combustion zone is very important. As the counter-flow geometry is more suited to support this

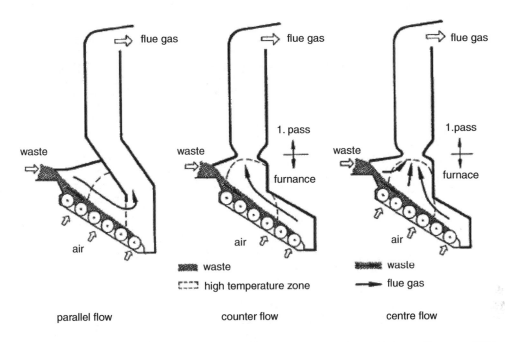

Figure 2.4 Schematic of Furnace designs for MSW incineration plants (Source: Goerner 2001; Reproduced with the permission of Prof. Dr. Klaus Goerner).

heat transfer, it is appropriate for higher LCV waste. The centre flow is good for a range of LCV (Fig. 2.4).

Each of these designs has advantages as well as disadvantages (Table 2.1). Several variants in furnace geometries deviating from these standard concepts are introduced by different manufacturers and are currently available.

Usually auxiliary firing systems, often fuelled by oil, are used to keep the combustion gases at the desirable temperature levels.

Boiler types

The mass burn boiler types can be divided into mass burn water wall, rotary water wall, and refractory wall designs. Water wall designs, presently in practice, have water-filled tubes in the furnace walls that are used to recover heat that produces steam. Rotary water wall combustors use a rotary combustion chamber constructed of water-filled tubes followed by a water wall furnace. Mass burn refractory designs are older, used in early 1970s primarily for reducing the volumes of MSW before sending to landfills and typically do not include any heat (energy) recovery.

Normally water wall boilers are most often have four passes: 3 vertical radiation passes and a convection pass. The first of the radiation passes is integrated in the furnace as the post combustion chamber. The convection pass, in which the evaporators, superheaters and economizers are located, may be vertical or horizontal. Typical designs of waste mass burn facilities utilize horizontal boiler (e.g., Hitachi Zosen Inova) and vertical boiler (e.g., Martin GmbH) configurations. Both the boiler configurations

Table 2.1 Advantages and disadvantages of different furnace geometries.

	Parallel flow	*Counter flow*	*Centre flow*
Advantages	Pyrolysis gases pass the hottest area and are burnt out; Suitable for lower LCV	Energy transfer from main combustion area to the drying and gasification area; Suitable for higher LCV	Very flexible for different heat release distributions on the grate; Suitable for a wider range of LCV
Disadvantages	Energy transfer from main combustion area to ignition area are just by radiation	Pyrolysis gases can bypass the hottest area and may cause burnout problems	Flow and mixing pattern after passage to the 1. pass sensible to disturbances

(Sources: Gorner 2011, Quina et al. 2011)

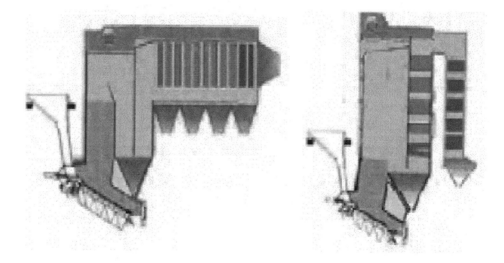

Figure 2.5   Schematic of Boiler configurations: (a) horizontal, (b) vertical (EC 2006).

(Fig. 2.5a) meet the state-of-the-art design and have many plants operating worldwide, the main difference being that horizontal boiler configuration requires more site space.

In both the systems, a number of empty passes with evaporation walls are followed by heat transfer surfaces i.e. evaporator, superheater and economiser. In the horizontal configuration, the flue gas in the convective heating surfaces travels horizontally. The horizontal boiler employs mechanical rapping tube cleaning methods requiring low internal steam consumption, whereas vertical boiler configuration employs soot blowers requiring higher steam consumption.

In the horizontal design, the dirt from the convective heating surface cleaning enters the hoppers without passing other heating surfaces on its way, thus reducing the risk of blocking the tube bundles resulting in a better availability. Another benefit of the horizontal design is that the support for the heating surfaces can be placed outside the flue gas.

In vertical configuration, the flue gas in the convective heating surfaces travels in the vertical direction; and the convective heating surfaces are usually cleaned by soot blowers which is very effective and minimizes the risk of blocking of the tube bundles. To avoid soot blower-induced erosion, hot tubes (i.e. superheater and evaporator tubes) are protected by stainless steel tube shells. Another advantage of this type is that very compact boilers can be designed because the vertical heating surfaces can use a common hopper for ash extraction, thus optimizing performance per ton of steel in the boiler. The arrangement of the tubes is such that the tube bundles do not need separate drains, a major advantage in terms of the time required for replacing the bundles (Lisa Branchini 2012). The investment cost for the vertical boiler structure is comparatively less.

The technical design features of boilers used in waste incineration and how to select low emission boilers are available from published literature (e.g., Walter Niessen 2002, Oland 2002; Babcock Walcox Volund Brochure).

Corrosion problem: When designing a boiler for waste incineration it is important to examine the risk of corrosion arising mainly due to the presence of corrosive substances such as alkalines, chlorine, sulfur, and heavy metals, zinc and lead, to an extent of nearly 1% (Hunsinger 2010, Ma and Rotter 2006, Albina 2005, Themelis 2005, Rademarkers et al. 2002, Yokoyama et al. 2001, Agarwal and Grossmann 2001, Covino et al.). During combustion of waste, these are released as HCl, $NaCl$, KCl, $ZnCl_2$, $PbCl_2$ and so on. These contribute to a highly corrosive atmosphere that shortens the life of the heat exchanger tubes in the water wall section due to the deposition, and especially in the steam superheater sections where the tube temperature is at its maximum. HCl is highly corrosive at temperatures $>450°C$ as well as at $<110°C$. Therefore, as a general accepted limit for a sufficient lifetime of boiler tubes, steam parameters of 40 bar/400°C are widely used. That is, the maximum steam temperature in the present WtE plants is an economic compromise between the acceptable corrosion rate and maximum power generation.

Actually, various types of corrosion processes exist in WTE power plants such as initial corrosion, high temperature corrosion, O_2-deficient corrosion, molten salt corrosion, chloride-high temperature corrosion, electrochemical corrosion, dew point corrosion and so on (Albina 2005, Lisa Branchini 2012). The corrosion process, although extensively studied during the last few years, it is not fully understood yet (Fleck 2012).

In addition to the concentration of corrosive content, particularly chlorine in the municipal waste, other factors that affect the rate of corrosion in the boilers are the operating temperature in the chamber, temperature fluctuations at a particular location that may disrupt the protective oxide layer, the method used for periodic cleaning of the process gas side of the tubes, and the design of the boiler (Rademarkers et al. 2002, Albina 2005).

To prevent the corrosive attacks on the furnace boiler system, the heating surfaces in the radiant part is protected by a resistant refractory material such as Al_2O_3, SiO_2, SiC, or Ni-based alloy materials or Inconel cladding (Wilson et al. 1997, Solomon 1998, Fukuda et al. 2000, Kawahara et al. 2002, Kawahara 2002, Spiegel 2002,

Table 2.2   Residence times in the after-burning zone vs Temperature.

Flue gas temp. (°C)	Residence time	
	After-burning zone (sec)	Furnace + after-burning zone (sec)
>950	0.3	1.4
>900	1.25	2.35
>850	2.15	3.25
>800	3.3	4.4

(Source: Görner 2011)

Uusitalo et al. 2002, Higuera et al. 2002, Al-Fadhli et al. 2002, Zwahr 2003, Hunsinger 2010). Also, by controlling process conditions such as (i) improvement of process control, particularly, by minimizing fluctuations in gas temperature and (ii) design modifications, to alter flow dynamics, enhancing mixing of gas through gas recirculation, and design of the boiler system. In addition to protecting from chemical reactions, thermal insulation is also crucial. The heat extraction from the furnace is restricted by adjusting the thickness of the material layer and by varying the thermal conductivity coefficient of the material.

The selection of the lining material is very critical for process engineering because the residence times in the after-burning zones are strongly dependent on the temperature levels which decide the production of hydrocarbons and CO as already explained. The residence times as a function of the temperature are given in Table 2.2.

In the radiant passes the flue gas is cooled slowly to a temperature of less than 700°C before it is further cooled by the heating surface bundles in the convections pass. To prevent low temperature corrosion, the feed water should be preheated to minimum 125°C before being introduced in the boiler (BW Volund brochure).

Energy recovery

In modern thermal incinerators, the chemical energy of the waste is finally released into the flue gases of the combustion process; therefore, the flue gases are at temperature of around 1000°C. The heat in the flue gas is recovered in the boiler, superheater and economizer. The heat turns the water in the boiler tubing to superheated steam having temperatures typically, 400°C at a pressure of 40 bars for generating electricity.

As explained earlier, the steam parameters are limited by the corrosive contents of flue gases. Therefore, higher temperatures in the superheater can be driven only with special measures and interesting concepts as being done in the waste-to-energy plants at Amsterdam (The Netherlands), Mainz (Germany) and Bilbao (Spain). These plants are described under Case studies in Chapter 5.

The recovered heat can be utilized in either of three ways depending on the boiler technology: (a) direct use as heat for district heating (usually through a 'steam cycle') or as process steam to industry; or (b) conversion to electricity using steam turbines; or (c) combined heat and electricity (CHP), i.e., utilizing directly as heat and also by generating electricity. A part of the generated energy – depending on the process technology – is often used internally in the operation of the plant.

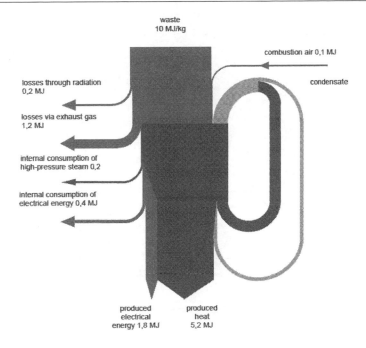

waste
10 MJ/kg

combustion air 0,1 MJ

condensate

losses through radiation
0,2 MJ

losses via exhaust gas
1,2 MJ

internal consumption of
high-pressure steam 0,2

internal consumption of
electrical energy 0,4 MJ

produced
electrical
energy 1,8 MJ

produced
heat
5,2 MJ

Figure 2.6a Energy flow chart diagram of pure power generation with normal steam parameter [Source: Federal Ministry of Agriculture and Forestry, Environment and Water Management, Vienna, Austria: State of the Art for Waste Incineration Plants, November 2002; Authors: Josef Stubenvoll (TBU), Siegmund Böhmer (UBA), Ilona Szednyj (UBA); Coordination of the joint study: Gabriele Zehetner].

The overall efficiency of a waste incineration plant is defined as the ratio of utilizable energy (heat and power) to supplied energy. If waste heat is fully used (CHP) a theoretical overall efficiency up to 80% can be achieved. When applying flue gas condensation, also the latent heat can be recovered, raising that figure to well over 90%. If heat is not used the overall efficiency applying normal steam parameters will be about 20%. Typical energy flow charts, assuming the feed waste calorific value as 10 MJ/kg, in the two cases are shown in Figures 2.6a & b.

The energy recovery from 1 ton of waste incineration is in the range of 400–700 kWh of electricity and additional 1205 kWh as heat (BREF 2006) with local variations. Typical mass burn facilities have energy recovery efficiencies of 14% to 27%, assuming that the energy from combustion is being converted into electricity (AECOM Canada Ltd. 2009).

Maximizing Efficiency/Energy recovery: In conventional modern plants electrical efficiency is usually limited to around 22–25% (gross) because the steam conditions in the plant have typically been restricted to 400°C, 40 bar. This restriction is to avoid serious corrosion problems resulting as explained earlier.

A new combustion concept was developed at Institute of Technical Chemistry, KIT, Germany, to generate very high temperature steam without corrosion and fouling problems (Hunsinger 2010). In this, a small part of the primary formed fuel gas is extracted before entering the flue gas burnout zone (Fig. 2.2b). This fuel gas passes a small bypass system equipped with the subsequently arranged process steps, gas

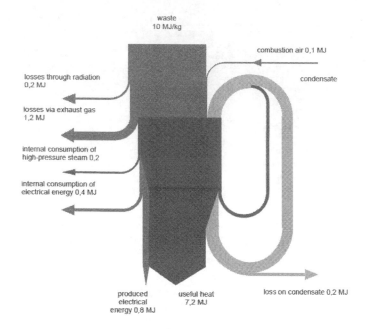

waste
10 MJ/kg

combustion air 0,1 MJ

condensate

losses through radiation
0,2 MJ

losses via exhaust gas
1,2 MJ

internal consumption of
high-pressure steam 0,2

internal consumption of
electrical energy 0,4 MJ

produced
electrical
energy 0,8 MJ

useful heat
7,2 MJ

loss on condensate 0,2 MJ

Figure 2.6b   Energy flow chart diagram of cogeneration (CHP) with normal steam parameters [Source: Federal Ministry of Agriculture and Forestry, Environment and Water Management, Vienna, Austria: State of the Art for Waste Incineration Plants, November 2002; Authors: Josef Stubenvoll (TBU), Siegmund Böhmer (UBA), Ilona Szednyj (UBA); Coordination of the joint study: Gabriele Zehetner].

cleaning, combustion and further superheating of the pre-generated steam from the main boiler. The off-gas and the removed solid residues are recycled back to defined locations in the main furnace. This process combines the advantages of combustion and gasification and allows to generate very high temperature steam. This principle also helps to lower NO_x generation and ensures low stack emissions and minimum amount of high quality residues that can be utilized or landfilled. The experimental details and results are discussed in the paper cited.

In recent decades, in Europe and the U.S. a significant number of plants have been operating using high pressure boilers and taking other measures to increase the overall thermal efficiency (4th Generation in incineration). For large-scale moving grate combustion systems, the following technical approaches are currently employed (WSP 2013):

(i)   Advanced combustion control: The enhanced process control will maximise combustion efficiency to ensure maximum burn-out of the organic waste content and decrease of excess air levels; and using FGR, optimum oxygen levels can be achieved;

(ii) High steam pressure and superheat temperature: Higher steam pressure and temperature will increase the enthalpy of the steam and allow greater energy to be recovered in the steam turbine. The boiler design to withstand such temperatures and pressures needs extreme care to protect the superheaters and increase the overall thermal efficiency. By locating the superheater tubes in the furnace, the steam temperatures can be enhanced beyond the ordinarily possible temperatures. To protect the tubes (Inconel) and to avoid major corrosion problems, they are located behind protective walls;

(iii)   Reheat cycle: Using a reheat cycle can increase the efficiency by several percent. Steam from the outlet of the high pressure stage of the turbine is sent back to the boiler where it is heated back to the original temperature, before being expanded in the low-pressure stage. This is a relatively high cost approach, and requires a careful evaluation of the balance between the cost and the benefit of increased electricity generation;

(iv)   Reduced boiler exit temperature: The boiler exit temperature is established by sizing the economizer; and is normally set much above the dew points for acid gases to reduce corrosion, and the dew point of moisture to prevent agglomeration of particulate on the boiler tubes. However, keeping the temperature well above the dew points will reduce energy recovery from the flue gases. Utilizing careful controls and providing extra corrosion protection in the economizers, some recent plants have tried to maximize the energy recovery at reduced temperatures;

(v)   Reduced steam condenser pressure: The condenser temperature has a strong influence on the plant efficiency, the lower the condenser temperature, the greater the pressure drop across the turbine which increases power generation. Water cooled condensers can create the lowest temperatures compared to air cooled condensers. However, if warmer water is used, it may not result in significant improvement in power cycle efficiency and will not offset the increased maintenance effort of a pumped once-through ocean water cooling system;

(vi)   Integration with fossil fuel-fired power plant (external superheating): If the plant is integrated with a gas turbine (CCGT) (*combined cycle gas turbine*) system using the high temperature exhaust gases from the gas turbine then additional heat can be provided. This can help boost the efficiency of energy recovery from the waste incinerator;

(vii)   CHP operation: The recovery of both heat as well as electricity can produce the highest increase in efficiency. Steam can be extracted from the turbine and used directly for process heating in industry or used to produce hot water for a district heating.

All of the above measures, while improve efficiency, involve considerable costs. Hence a balance has to be evaluated between additional capital and operational costs and the increased revenue from the sales of power and/or heat.

There are a good number of Plants operating at high steam temperatures and pressures in Europe – Italy, Germany, France, The Netherlands, Denmark, and England – utilizing some of these approaches.

Current status

The Moving Grate technology has been the most preferred technology for over a century and more than 1000 plants operate globally (Santec Final report 2011). Considerable progress has been made in the development with regard to the old processes with grate systems. Optimization of the design of the combustion chamber,

CFD simulations, flue gas recirculation, enrichment of the primary air with oxygen, water-cooled grate elements, and advances in the combustion control (e.g. IR-camera, Acoustic gas temperature measurement, etc.) are the new developments in the grate systems (see Beckmann and Scholz 2002). To make these technologies more widespread and popular, apart from upgrading the technical aspects, the public awareness must be enhanced, and the technology should be made environment-friendly through strict legislative measures.

Scaling up mass burn facilities

Mass-burn facilities can be scaled in capacity anywhere from approximately 36,500 to 365,000 tons/year (tpy) per operating unit (GENIVAR Ontario Inc. 2007, AECOM Canada Ltd. 2009). These facilities generally consist of multiple modules or furnaces and can be expanded through addition of more units and supporting ancillary infrastructure as required. Multiple modules can often be accommodated on a single site with some sharing of infrastructure such as tip floor, ash management areas, and stack.

Modular, Two Stage Combustion: Modular combustors are similar to mass burn combustors burning waste that is not pre-processed, but they are typically fabricated and generally smaller in size, from 4 to 130 TPD. One of the most common types of modular combustors is the deficit-air or controlled-air type. They are used where start-ups occur each day and/or where throughputs are low, for example at commercial/factory sites or in rural areas.

(i)   Modular Starved-air Combustors: The combustor consists of two combustion chambers, 'primary' and 'secondary' chambers. Waste is fed to the primary chamber by a hydraulically activated ram. The charging bin is filled by a front end loader or other means. Waste is fed automatically at regular intervals, generally with 6 to 10 minutes gap, and moved through the primary combustion chamber by either hydraulic transfer rams or reciprocating grates.

Combustors using transfer rams have individual hearths upon which combustion takes place. Grate systems generally include two separate grate sections. In either case, waste retention times in the primary chamber are long, lasting up to 12 hours. Bottom ash is usually discharged to a wet quench pit and managed as done in grate incinerators. Air is introduced into the primary combustion chamber at sub-stoichiometric levels, resulting in a flue gas rich in unburned hydrocarbons. The quantity of air introduced into the primary chamber defines the rate at which waste burns. The air flow rate to the primary chamber is controlled to maintain an exhaust gas temperature generally at 650 to 980°C (1200 to 1800°F), which corresponds to about 40 to 60% of the required air.

The hot, fuel-rich flue gases flow to the secondary chamber, and are mixed with oxygen-rich air to complete the burning (oxidation) process. The air quantity added to the secondary chamber is controlled to maintain a desired flue gas exit temperature, typically 980 to 1200°C (1800 to 2200°F). Approximately 80% of the total combustion air is introduced as secondary air; and typical excess air levels vary from 80 to 150%.

The walls of both combustion chambers are refractory lined. The heat generated in the secondary chamber is fed into a heat recovery boiler. Early designs did not include energy recovery, but newer installations have heat recovery boilers. At times, two or more combustion modules manifold to a single boiler. Most of these combustors are equipped with auxiliary fuel burners located in both the primary and secondary combustion chambers. Auxiliary fuel can be used during startup (many modular units do not operate continuously) or when problems are experienced maintaining desired combustion temperatures. In general, the combustion process is self-sustaining through control of air flow and feed rate, so that continuous co-firing of auxiliary fuel is normally not necessary. The high combustion temperatures and proper mixing of flue gas with air in the secondary chamber provide good combustion, resulting in relatively low CO and trace organic emissions.

As the amount of air introduced through the primary combustion chamber is limited, gas velocities in the primary chamber and the amount of entrained particulate matter are low. As a result, PM emissions from modular starved-air combustors are relatively low. Many existing smaller systems do not have air pollution controls. A few of the newer systems have acid gas/PM controls.

(ii)   Modular Excess Air Combustors: There are fewer of this type compared to the earlier one. The design is similar to that of controlled-air type. Waste is batch-fed to the refractory-lined primary chamber and is moved through the chamber by hydraulic transfer rams, oscillating grates, or a revolving hearth. Bottom ash is discharged to a wet quench pit. Additional flue gas residence time for fuel/carbon burnout is provided in the secondary chamber, which is also refractory-lined. The heat from flue gases is typically recovered in a waste heat boiler.

Facilities with multiple combustors may have a tertiary chamber where flue gases from each combustor are mixed prior to entering the energy recovery boiler.

Unlike the controlled-air combustor, the excess-air combustor typically operates at about 100% excess air in the primary chamber, but may vary between 50 and 250 percent excess air. These combustors also use FGR to maintain desired temperatures in the primary and secondary chambers. Due to higher air velocities, PM emissions from these type combustors are higher than those from the first type, and more similar to the levels from mass burn units. However, NO_x emissions from these combustors appear to be lower than from either controlled-air type or mass burn units.

2.6.2   Fluidized bed combustion technology

An alternative to grate combustion systems are the fluidized bed incinerator (Howe and Divilio 1993, Legros 1993, Patel and Wheeler 1994, Rhyner et al. 1995). In fluidized bed combustion, waste fuel requires pretreatment, i.e., shredding, sorting and removing noncombustible materials in order to generate a more homogenous solid fuel. The fuel is fed into a combustion chamber through ports located on chamber wall, in which there is a bed of inert material (usually sand) in a fluid state on a grate or distribution plate.

The thermal destruction of waste in a fluidized-bed incinerator requires essentially the same sequence for progressive destruction of the material as in grate technology, but instead of occurring in discrete zones, all of the processes occur simultaneously in a single large bed. Air and small particles of waste are continuously injected at a high rate into the bed, and sufficient oxygen is always available to all parts of the bed. This allows each distinct piece of waste to undergo its drying, volatilization, oxidation of the gases and vapors, combustion of organic solids, and complete burn-down of the char in all parts of the bed at the same time.

The combustion gases are retained in a combustion zone above the bed. The heat from combustion is recovered by devices located either in the bed or at the exit point of the gases from the chamber or at both locations.

The surplus ash is removed at the bottom of the chamber and managed in the same way as bottom ash from a moving grate system (mass burn incineration). The recovered metal fraction is sent for recycling.

Fluidized bed systems operate in a temperature range, 750–850°C. As they operate with only 30–40% excess air, they are more efficient in energy recovery than grate furnaces (Bontoux 1999). The boilers are of two types: the Bubbling fluidized bed (BFB) type and the Circulating fluidized bed (CFB) type as utilized in coal combustion also (Jayarama Reddy 2014).

In the bubbling fluidized bed combustion, the airflow is just sufficient to mobilise the bed and provide good contact with the waste. The airflow is not high enough to allow large amounts of solids to be carried out of the combustion chamber. In the circulating fluidized bed, on the other hand, the velocity of the air is much higher which results in a circulation of bed material, ash and burning fuel in a loop consisting of the combustion chamber and a primary cyclone. The differences in the relationship between air flow and bed material have implications for the type of wastes that can be burned, as well as the heat transfer to the energy recovery system. The Hitachi Zosen Inova's BFB and CFB incinerators are schematically shown in Figure 2.7a & b.

The advantage of FB technology is its ability to burn high moisture waste efficiently and high heat flux per square meter of combustion chamber cross section. Further, in fluidized-bed systems, heat transfer from the sand bed to the heat exchanger walls is highly efficient, NO_x and SO_x emissions are less, and the amount of unburned matter in the ash residues is often low (Abbas et al. 2003). The drawback of the process is a larger portion of fly ash is generated (6% compared to 2% for mass burn units) due to the particulate present in the fluidized bed itself. It is believed, by altering the cyclone configuration after the combustion chamber, this can be reduced.

Theoretically, a fluidized bed may be applied for combustion of pretreated MSW, as the technology has a number of appealing features as incineration technology. However, the merits are not yet thoroughly proven on municipal solid waste. The fluidized bed may be good for special types of industrial waste, and for this purpose it is widely functional in Japan (World Bank 1999a). In terms of emissions, modern plants can reduce their heavy metal output below European and North American standards (Oka 2003). In general, fluidized-bed incinerators appear to operate efficiently on smaller scales than do mass-burn incinerators.

There is renewed interest in the FBC technology in Europe with increasing amounts of high calorific MSW which is more difficult to combust in grate systems. Around 450 WtE facilities operate in Europe, and 30 of them utilize fluidized bed technology. Most of these use a feed stock mixture of MSW, sewage sludge, industrial waste, pre-sorted organic waste, RDF or woodchips. Very few facilities are using *only* MSW. In UK, the fluidized bed technology is less popular for combusting MSW but widely

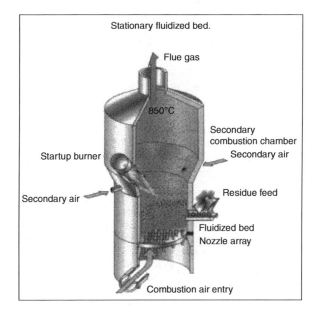

Figure 2.7a Schematic of Hitachi Zosen Inova's Bubbling Fluidized Bed (Source: HZ-Inova Brochure; reproduced with the permission of Hitachi Zosen Inova AG – Dr. Michael Keunecke).

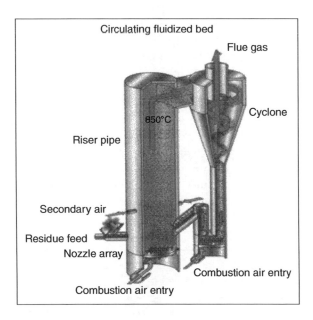

Figure 2.7b Schematic of Hitachi Zosen Inova's Circulating FB (Source: HZ-Inova Brochure; reproduced with the permission of Hitachi Zosen Inova AG – Dr. Michael Keunecke).

applied to sewage sludge (Defra 2013). In China, it is an emerging technology for waste incineration and since beginning of 21st century, 47 CFB plants have been built with a capacity to treat 14.6 million tons of MSW (Huang et al. 2013).

2.6.3 Refuse-derived fuel combustion

RDF is mechanically processed MSW which is easily storable, transportable, and more homogeneous fuel for combustion (Jayarama Reddy 2011). Although the number of RDF systems growing, they represent a much smaller share of traditional incineration facilities. RDF technologies were discussed in detail by Shepherd and Gupta of NREL, USA (1992).

The type of RDF used is dependent on the boiler design. It usually utilizes spreader stokers and fire fluff RDF in a semi-suspension mode. Pelletized RDF is fed into the combustor through a feed chute using air-swept distributors; this allows a portion of the feed to burn in suspension and the remainder to burn out after falling on a horizontal traveling grate. The traveling grate moves from the rear to the front of the furnace, and distributor settings are adjusted so that most of the waste settles on the rear two-thirds of the grate. This allows more time to complete combustion on the grate. Underfire and overfire air are introduced to enhance combustion, and these incinerators typically operate at 80 to 100% excess air. Waterwall tubes, a superheater, and an economizer are used to recover heat for production of steam or electricity. The 1995 inventory indicated that dedicated RDF facilities range from 227 to 2,720 metric tons per day total combustion capacity.

Pulverized coal-fired boilers can co-fire fluff RDF or powdered RDF. Due to its high moisture content and large particle size, RDF requires a longer burnout time than coal. RDF can also be co-fired with coal in stoker. Both the designs of fluidized-bed incinerator also burn RDF. Waste-fired fluidized-bed RDFs typically operate at 30 to 100% excess air levels and at bed temperatures around 815°C. The 1995 inventory indicated that fluidized-bed municipal waste combustors have capacities ranging from 184 to 920 metric TPD. These systems are usually equipped with boilers to produce steam.

RDF systems have two basic components: RDF production and RDF incineration. In the production facilities, RDF is prepared in various forms through materials separation, size reduction, and pelletizing. Although RDF processing has the advantage of removing recyclables and contaminants from the combustion stream, the complexity of this processing results in high operating and maintenance costs and reduced reliability of RDF production facility. On average, capital costs per ton of capacity for incineration units that use RDF are higher compared to other incineration options.

RDF production plants also have an indoor tipping floor. The waste is typically fed onto a conveyor, which is either below grade or hopper fed. In some plants, the corrugated and bulky items are separated at the feeding point. The waste on the conveyor travels through a number of processing stages tailored to the desired products, beginning with magnetic separation, typically including one or more screening stages, trommel or vibrating screens, shredding or hammer milling of waste with additional screening steps, pelletizing or baling of burnables. And depending on the local recycling markets and the design of the facility, may include a manual separation line (see Jayarama Reddy 2010).

RDF in Europe, produced usually in the form of pellets or baled paper and plastic, is sold to electricity generating stations that use fluidized-bed technology. RDF in Europe is also burned for generating heat needed in industrial processes, particularly papermaking, and also co-combusted in utility generating stations designed for coal or wood.

Emissions from RDF-fired systems

Several publications on emissions from RFD-fired plants are found in the literature. The U.S. EPA has studied the emissions for RDF co-combustion with coal in a small spreader-stoker fired boiler, varying both the RDF type and amount of coal co-burned with the RDF. In two experiments a waste chemical, triethanolamine, was added to the fuel, and its destruction efficiency was assessed. Analysis of the flue gases identified low levels of hydrocarbons, NO_x, CO, and SO_2. The particulate loadings increased as the percentage of RDF in the total fuel increased. More than half of the particulate loading was submicron in size when RDF was fired without coal. Large quantities of POM's were detected in those experiments in which the shredded and pelletized RDF was fired in the furnace. No dioxins were detected in these tests. In the experiments with only RDF, lead emissions were several orders of magnitude above the levels detected in the coal base run. Based upon the flue gas analysis, the destruction of the triethanolamine was complete (Rising and Allen 2002, revised 2004).

Rotary kiln incinerator

Incineration in a rotary kiln is normally a two stage process consisting of a kiln and separate secondary combustion chamber. The kiln is a long cylindrical, refractory-lined, horizontal steel shell (the primary combustion chamber) and is inclined downwards from the feed entry point. The rotation around its longitudinal axis moves the waste through the kiln with a tumbling action which exposes the waste to heat and oxygen (Buonicore et al. 1992). To allow it to rotate, the kiln is fabricated with reinforced steel bands on the outside of the cylinder and these rings ride on steel rollers. Kilns are typically rotated by a gear train that engages a spur gear affixed to the circumference of the shell. The inner refractory lining serves to protect the kiln structure and needs to be replaced often. In the primary chamber, there is conversion of solid to gases, through volatilization, destructive distillation and partial combustion reactions. The secondary chamber is necessary to complete gas phase combustion reactions.

Most kilns for waste combustion applications are 4.5 to 6 metres in diameter and their length can range from twice to ten times their diameter depending upon the specific application. Some designs incorporate vanes or paddles to the inside surface to encourage solids mixing along the kiln length. Rotational speeds range from 0.5 to 2 revolutions per minute, again depending upon the nature of the wastes being handled. It is important to realize that the combination of the rake and the rotational speed combine to determine the residence time of the solids in the system and the amount of mixing provided for wastes and combustion air in the furnace. Mixing also serves to transfer heat between the waste, the flames and the refractory.

Rotary kilns used for MSW processing can be of water wall or refractory wall design; and Volund rotary furnace, and the Westinghouse/O'Connor rotary kiln are the systems currently in operation in Europe. The Westinghouse unit is a waterwall lined kiln with air and water tubes arranged longitudinally in the kiln whereas the Volund is a refractory lined kiln. There is also a system which oscillates a rotary kiln for smaller scale incineration of MSW with energy recovery.

This technology is not commonly used for mixed MSW; those under use have unit capacity range, 100 to 300 TPD. But it is *mainly* used for treatment of *hazardous waste and sewage sludge. Hazardous waste rotary kiln* systems are typically sized for 60 million Btu/hr heat input, but can be as large as 150 million Btu/hr. Solids retention times in the kiln range from 0.5 to 1.5 hours, while gasses are retained for approximately 2 seconds in most systems. Typical gas temperatures in the kiln exceed 870°C, while the solids attain temperatures in excess of 650°C. Combustion air is provided through ports on the face of the kiln, and through leakage through the rotary seals. The resulting excess air levels range from 50 to 200%.

In the rotary kilns, their seals are generally weak and cause for emissions. High rates of volatilisation can cause the pressure to increase in the kiln enabling the gasses to escape through the seals. Thus, the feeding of highly combustible/explosive wastes needs to be carefully controlled, or the rate of mixing must be reduced to minimise the exposure of fresh surfaces.

Inorganic materials, including ash, slag and other incombustible items that remain at the end of the kiln are discharged into a water-quench tank. The water acts as a seal preventing the air enter into the kiln. The rate of discharge of ash residue must be so controlled that hot masses do not drop into the quench tank and create steam explosions resulting in increased pressure in the kiln.

The gases from the kiln move into a secondary refractory lined combustion chamber for complete destruction of volatile gas phase unburned materials. This is essentially a liquid injection furnace. Temperatures of around 1200°C with 100 to 200% excess air, turbulent flow mixing, and a gas residence time of 1 to 3 seconds ensure suitable performance from the afterburner (secondary chamber). Temperatures are maintained through the use of auxiliary fuel, normally a liquid hazardous waste. Since slag can be a concern at the high temperatures in the afterburner, some facilities position hot cyclones between the kiln and the afterburner to remove entrained particulate matter. A schematic of rotary kiln for hazardous waste incineration is shown in Figure 2.8.

The process characteristics of the three types of combustion: grate incineration, fluidized bed and rotary kiln are summarized in Table 2.3 (Bosmans et al. 2013).

Key criteria for selecting incineration technology

The World Bank report (1999a) – Decision maker's guide to MSW Incineration – lists out the following 'mandatory' and 'strongly advisable' criteria:

Mandatory criteria

(a)   The technology should be based on the mass burning principle with a movable grate, and the supplier must have several plants successfully operating for a number of years;

(b)   The furnace must be designed for stable and continuous operation and complete burnout of the waste and flue gases ($CO < 50$ mg/Nm^3, TOC < mg/Nm^3);

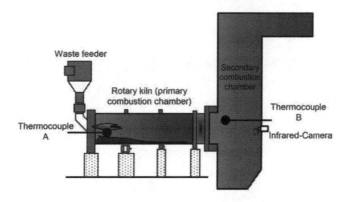

Figure 2.8 Schematic of Hazardous waste incineration system using a rotary kiln as primary combustor (Source: Du J. et al. 2012; reproduced with permission from Professor Qunxing Huang, Zhejiang University).

Table 2.3 Process characteristics of the three main incinerator types.

	Grate incinerator	*Rotary kiln*	*Fluidized bed*
Process description	The grate moves the waste through the various zones of the combustion chamber	Cylindrical vessel located on rollers which allow the kiln to rotate/oscillate around its axis, waste is conveyed by gravity	Lined combustion chamber in the form of a vertical cylinder, the lower section consists of a bed of inert material which is fluidized with air, waste is fed continuously into the fluidized sand bed
Used for	Mixed municipal wastes, possible additions; commercial and industrial non-hazardous wastes, sewage sludge, clinical wastes	Hazardous and clinical wastes	Finely divided wastes (e.g., RDF, sewage sludge)
Process temperature	850–1100°C	850–1300°C	Freeboard: 850–950°C Bed: 650°C or higher
Remarks	Most widely applied	i. very robust, allows the combustion of solid, liquid, gaseous wastes & sludges ii. a post-combustion chamber is added to enhance the destruction of toxic compounds	i. Bubbling type, used for sludges, ii. circulating type, used for dried sludge with high calorific value, iii. rotating type, allows for a wide range of calorific value fuels (e.g., co-combustion of sludges and pretreated wastes)

(Source: Bosmans et al. 2013)

(c)   The flue gases from the furnace must be cooled to 200°C or lower before being treated;

(d)   The flue gas cleaning equipment must be at least a two-field ESP (basic emission control, dust $<30\,mg/Nm^3$);

(e)   An organized landfill must be available for residue disposal where a full leachate control must be exercised.

Strongly advisable

(a)   The annual waste supply for incineration should not be less than 50,000 metric tons and the weekly variations in the supply to the plant should not exceed 20 percent.

(b)   Municipal solid waste incineration plants should be in locations dedicated to medium or heavy industry.

(c)   The stack should be twice the height of the tallest building within one kilometer, or at least 70 meters high.

2.6.4   Suppliers of the facilities

Conventional combustion (specifically mass burn) technology is well established, with a number of established vendors that supply some or all components of the technology. Based on a recent review, over 20 vendors worldwide were found to provide some components (grate systems, boilers) or provide services for the overall Design, Build and Operation (DBO) of conventional combustion facilities (WSP 2013).

In Europe, the four main suppliers of grates and potentially other components of mass burn incineration technology, who are also the primary suppliers of Grates in North America and Asia, are Babcock & Wilcox Vølund (Denmark), Fisia Babcock Environment GmbH (Germany), Martin GmbH (Germany), and Von Roll Inova (Switzerland). In Asia, Keppel Seghers have also supplied several grate fired plants.

REFERENCES

Abbas, Z., Moghaddam, A.P. & Steenari, B.-M. (2003) Release of salts from municipal waste combustion residues. *Waste Management*, 23, 291–305.

Achternbosch, M. & Richers, U. (2002) Materials Flows and Investment Costs of Flue Gas Cleaning Systems of Municipal Solid Waste Incinerators. Karlsruhe, Forschungzentrum Karlsruhe Wissenschaftliche Berichte (FZKA).

AECOM Canada Ltd. (2009) *Management of Municipal Solid Waste in Metro Vancouver—A Comparative Analysis of Options for Management of Waste After Recycling.*

Agarwal, D.C. & Grossmann, G.K. (March 2001) *Case Histories on the Use of Nickel Alloys in Municipal and Hazardous Waste Fueled Facilities, Corrosion.* Houston, TX, NACE International.

A.J. Chandler & Associates Ltd. (December 15, 2006) *Review of Dioxins and Furans from Incineration in Support of a Canada-Wide Standard Review.* A report prepared for The Dioxins and Furans Incineration Review Group, Canadian Council of Ministries of the Environment Inc.

Albina, D.O. (2005) *Theory and Experience on Corrosion of Waterwall and Superheater Tubes of Waste-to-Energy Facilities.* Master Thesis. New York, NY, Columbia University.

Al-Fadhli, H.Y., Stokes, J., Hashmi, M.S.J. & Yilbas, B.S. (2006) The erosion-corrosion behaviour of high velocity oxy-fuel (HVOF) thermally sprayed inconel-625 coatings on different metallic surfaces. *Surface and Coatings Technology*, 200 (20–21), 5782–5788.

Annepu, R.K. (January 2012) *Sustainable Solid Waste Management in India*. Thesis submitted for MS degree. New York, NY, Earth Resources Engineering Center, Department of Earth and Environment Engineering, Columbia University.

Babcock & Wilcox Volund. *21st Century Advanced Concept for Waste-Fired Power Plants*.

Bardi, S. & Astolfi, A. (December 2010) Modelling and control of a Waste-to-Energy plant—Waste bed temperature regulation. *IEEE Control Systems Magazine*, 27–37.

Baxter, D. & El Asri, R. (2004) Process control in municipal solid waste incinerators: Survey and assessment. *Waste Management & Research*, 22, 177–185.

Beckmann, M. & Scholz, R. (2002) Conventional thermal treatment methods. In: Ludwig, Ch., Hellweg, S. & Stucki, S. *Municipal Solid Waste Management—Strategies and Technologies for Sustainable Solutions*. Berlin – Heidelberg – New York, Springer-Verlag.

Bilitewski, B., Härdtle, G. & Marek, K. (1997) *Waste Management*. Berlin, Springer. ISBN: 3-540-59210-5.

Bontoux, L. (1999) *The Incineration of Waste in Europe: Issues and Perspectives*. European Commission Joint Research Centre.

Bosmans, A., Vanderrey, I., Geysenc, D. & Helsena, L. (2012) The crucial role of Waste-to-Energy technologies in enhanced landfill mining: A technology review. *Journal of Cleaner Production*, 1–14. doi:10.1016/j.jclepro.2012.05.032.

Branchini, L. 2012. Advanced Waste-to-Energy Cycles, Doctoral thesis, CIRI-ENA-Alma Mater Studiorum – Universita di Bologna, Rimini 47900, Italy.

BREF (2006) *Integrated Pollution Prevention and Control—Reference Document on the Best Available Techniques for Waste Incineration*. European Commission.

BREF Report (December 2001) Draft of a German Report with basic informations for a BREF-Document, 'Waste Incineration'. BREF Report English—FTP Direct Listing. Available from: files.gamta.It/aaa/Tipk/tipk/4_kiti%20GPGB/63.pdf.

Brunner, C.R. (1994) *Hazardous Waste Incineration*. 2nd edition. New York, NY, McGraw-Hill.

Buonicore, A., Davis, W. & Pakrasi, A. (1992) Combustion sources. In: Buonicore, A. & Davis, W. (eds.) *Air Pollution Engineering Manual*. Air & Waste Management Association, Van Nostrand Reinhold. p. 207.

Chapman, A.J. (1986) *Fundamentals of Heat Transfer*. Englewood Cliffs, NJ, Prentice-Hall.

Chen, D. (1995) Fuzzy logic control of batch-feeding refuse incineration. In: *Uncertainty Modeling and Analysis, Annual Conf. of the North American Fuzzy Information Processing Society; Proc. of ISUMA—NAFIPS '95; 17–19 Sept. 1995*. Available from: http://ieeexplore.ieee.org/xpl/articleDetails.jsp?arnumber=527669.

Cheng, H., Zhang, Y., Meng, A. & Li, Q. (2007) MSW fueled power generation in China: A case study of WtE in Changchun city. *Environmental Science & Technology*, 41, 7509–7515.

Clarke, M.J. (2002) Introduction to municipal solid waste incineration. In: *Air and Waste Management Association Annual Meeting, Baltimore, MD, June 23–27, 2002*.

Covino, B.S., Holcomb, G.R., Cramer, S.D., Bullard, S.J. & Ziomek-Moroz, M. *Corrosion in Temperature Gradient*. Available from: http://www.netl.doe.gov/publications/proceedings/03/materials/manuscripts/Holcomb_m.pdf.

DeFra (2007) *Incineration of Municipal Solid Waste*. Report of Department for Environment, Food & Rural Affairs, UK. Available from: www.gov.uk/.

DeFra (2013) *Incineration of Municipal Solid Waste*. Govt. of UK. Available from: https://www.gov.uk/government/uploads/system/uploads/attachment_data/file/221036/pb13889-incineration-municipal-waste.pdf.

Doing, M. (2013/2014) *Time Out—For Waste to Energy in Europe*. 14 (5). Available from: http://www.waste-management-world.com/articles/print/volume-14/issue-5/features/time-out-for-waste-to-energy-in-europe.html.

Du, J., Huang, Q. & Yan, J. (2012) Method for determining effective flame emissivity in a rotary kiln incinerator burning solid waste. *Journal of Zhejiang University-Science A (Applied Physics & Engineering)*, 13 (12), 969–978; (hqx@zju.edu.cn).

European Commission (August 2006) *Integrated Pollution Prevention and Control. Reference Document on the Best Available Techniques for Waste Incineration.*

European Union (November 19, 2008) Directive 2008/98/EC of the European Parliament and of the Council of on waste and repealing certain Directives.

Federal Ministry of Agriculture and Forestry, Environment and Water Management, Vienna, Austria (November 2002) *State of the Art for Waste Incineration Plants.* Authors: Stubenvoll, J. (TBU), Böhmer, S. (UBA), Szednyj, I. (UBA); Coordination of the joint study: Zehetner, G.

Fleck, E. (October 2012) *Waste Incineration in 21st Century—Energy Efficient and Climate-Friendly Recycling Plant & Pollutant Sink.*

Frey, H.H., Peters, B., Hunsinger, H. & Vehlow, J. (2003) Characterization of municipal solid waste combustion in a grate furnace. *Waste Management*, 23 (8), 689–701.

Fukuda, Y., Kawahara, K. & Hosoda, T. (2000) Application of high velocity flame sprayings for superheater tubes in waste incinerators. *Corrosion 2000*, p. 00264.1-00264.

GENIVAR Ontario Inc. in association with Ramboll Danmark A/S (2007) *Municipal Solid Waste Thermal Treatment in Canada.*

Gorner, K. (2001) Waste incineration—State-of-the-art and new developments. In: *IFRF 13th Members Conference, Noordwijkerhout, The Netherlands, 15–18 May 2001.*

Guyer, J.P. *Introduction to Solid Waste Incineration.* Course No.C04-024. Stony Point, NY, Continuing Education and Development, Inc. Available from: http://www.cedengineering.com/upload/Intro%20to%20Solid%20Waste%20Incineration.pdf.

Habeck-Tropfke, H. (1985) *Müll- und Abfalltechnik.* Dusseldörf, Werner Verlag.

Hickman Jr., L. (2003) *American Alchemy: The History of Solid Waste Management in the United States.* Santa Barbara, CA, Forester Press.

Higuera, V., Belzunce, F.J., Carriles, A. & Poveda, S. (2002) Influence of the thermal-spray procedure on the properties of a nickel-chromium coating. *Journal of Materials Science*, 37 (3), 649–654.

Hites, R.A. (2011) Dioxins: An overview and history. *Environmental Science & Technology*, 45, 16–20.

Howe, R.C. & Divilio, R.J. (1993) *Fluidized Bed Combustion of Alternative Fuels.* EPRI Report, TR-1000547s.

Huang, Q., Chi, Y. & Themelis, N.J. (2013) A rapidly emerging WtE technology: CFB combustion. In: *Proc. International Conf. Thermal treatment technologies (IT3), San Antonio, TX, October 2013.*

Hunsinger, H. (2010) A new technology for high efficient WTE plants. In: *2nd W2W & I-CIPEC Conference, 27–29 July 2010, Kuala Lumpur, Malaysia.*

IAWG (1997) Municipal solid waste incinerator residues. The international ash working group, In: Chandler, A.J., et al. (eds.) *Studies in Environmental Science.* Vol. 67. The Netherlands, Elsevier.

IEA Bioenergy *Accomplishments from IEA Bioenergy Task 36: Integrating Energy Recovery into Solid Waste Management Systems (2007–2009)—End of Task Report.* Chapter 4: Contributors—Beciden, M. (SINTEF), Vehlow, J. (KIT) & Howes, P. (AEA). Available from: www.ieabioenergytask36.org/.../2007-2009/Introduction_Final.pdf.

Jansson, S. (November 2008) *Thermal Formation and Chlorination of Dioxins and Dioxin-Like Compounds.* Doctoral Thesis. Sweden, Chemistry Department, Umea University.

Jayarama Reddy, P. (2011) *Municipal Solid Waste Management—Processes, Energy Recovery, Global Examples.* BS Publications, Hyderabad & CRC Press/Balkema, Leiden.

Jayarama Reddy, P. (2014) *Clean Coal Technologies for Power Generation*. The Netherlands, CRC Press/Balkema, Leiden. ISBN: 97-1-138-00020-9 (Hbk); ISBN: 978-0-203-76886-0 (eBook PDF).

Jenkins, B.M., Baxter, L.L. & Miles Jr., T.R. (1998) Combustion properties of biomass. *Fuel Processing Technology*, 54, 17–46.

Jungten, H., Richter, E., Knoblauch, K. & Hoang-Phou, T. (1988) Catalytic NO_x reduction by ammonia and carbon catalysts. *Chemical Engineering Science*, 43, 419–428.

Kalogirou, E., Themelis, N., Samaras, P., Karagiannidis, A. & Kontogiannis, S.T. *Fly Ash Characteristics from Waste-to-Energy Facilities and Processes for Ash Stabilization*. Available from: http://www.iswa.org/uploads/tx_iswaknowledgebase/Kalogirou.pdf [Retrieved 20th April 2015].

Kawahara, Y. (2002) High temperature corrosion mechanisms and effect of alloying elements for materials used in waste incineration environment. *Corrosion Science*, 44 (2), 223.

Kawahara, Y., Takahashi, K., Nakagawa, Y., Hososda, T. & Mizuko, T. (2000) Demonstration test of new corrosion-resistant superheater tubings in a high-efficiency Waste-to-Energy plant. *Corrosion 2000*, p. 265.

Kilgroe, J.D., Nelson, L.P., Schindler, P.J. & Lanier, W.S. (1990) Combustion control of organic emissions for MSW combustors. *Combustion Science and Technology*, 74, 223–244.

Kokalj, F. & Samec, N. (March 2013) Combustion of MSW for power production. In: Hoon Kiat Ng (ed.) *Advances in Internal Combustion Engines and Fuel technologies*. InTechOpen. pp. 277–309. ISBN: 978-953-51-1048-4.

Kreith, F. (1959) *Principles of Heat Transfer*. Scranton, PA, International Text Book Co.

Legros, R. (1993) *Energy from Waste—Review in the Field of FBC of MSW*. Report to CANMET under DSS contract 28SS.23440-1-9070.

Leskens, M., van Kessel, L.B.M. & Bosgra, O.H. (2005) Model predictive control as a tool for improving the process operation of MSW combustion plants. *Waste Management*, 25, 788–798.

Leskens, M., van der Linden, R.J.P., van Kessel, L.B.M., Bosgra, O.H. & Van den Hof, P.M.J. (2008) Nonlinear model predictive control of municipal solid waste combustion plants. In: *Intl. Workshop on Assessment and Future Directions of NMPC, Pavia, Italy, September 5–9, 2008*. Available from: www.dcsc.tudelft.nl/.../Paperfiles/Leskens&etal_NMPC08_Pavia.pdf.

Leskins, M., van Kessel, L.B.M., Van den Hof, P.M.J. & Bosgra, O.H. *Model Predictive Control of MSW Plants*. Available from: www.dcsc.tudelft.nl/Research/old/project_ml_pvdh_ob.html.

Li, Y.S., Niu, Y. & Wu, W.T. (2003) Accelerated corrosion of pure Fe, Ni, Cr and several Fe-based alloys induced by $ZnCl_2$–KCl at 450°C in oxidizing environment. *Materials Science and Engineering A*, 345, 64–71.

Liang, Z. & Ma, X. (2010) Mathematical modeling of MSW combustion and SNCR in a full-scale municipal incinerator and effects of grate speed and oxygen-enriched atmospheres on operating conditions. *Waste Management*, 30 (12), 2520–2529.

Ma, W. & Rotter, S. (May 2006) Overview on the chlorine origin of MSW and Cl-originated corrosion during MSW & RDF combustion process. In: *Proc. of the Second Intl. Conf. on Bioinformatics and Biomedical Engineering (iCBBE '08)*. pp. 4255–4258.

Matthes, J., Keller, H.B., Fouda, Ch. & Schreiner, R. (2004) On the optimization of industrial combustion processes using infrared thermography. In: *Proceedings of the 23rd IASTED Intl. Conf. Modelling, Identification and Control*. Vol. 23. pp. 386–391.

Miller, J. (August 2011) *Waste-to-Energy in U.S. in 2010*. Office of RCR, US EPA.

Morin, O. (October 2014) *Technical and Environmental Comparison of CFB and Moving Grate Reactors*. Master's Thesis in Earth Resources Engineering. New York, NY, Department of Earth & Environmental Engineering, Columbia University.

Moy, P., Krishnan, N., Ulloa, P., Cohen, S. & Brandt-Rauf, P.W. (2008) Options for management of MSW in New York City: A preliminary comparison of health risks and policy implications. *Journal of Environmental Management*, 87, 73–79.

Müller, B., Keller, H.B. & Kugele, E. (1998) Fuzzy control in thermal waste treatment. In: *Sixth European Congress on Intelligent Techniques and Soft Computing (EUFIT '98)*. Vol. 3. pp. 1497–1501.

Niessen, W.R. (2002) *Combustion and Incineration Processes: Applications in Environmental Engineering*. 3rd edition. New York, NY, Marcel Dekker Inc.

Oka, S. (September 2003) *Fluidized Bed Combustion*. Boca Raton, FL, CRC Press.

Oland, C.B. (2002) *Guide to Low Emission Boiler and Combustion Equipment Selection*. Prepared by Oak Ridge National laboratory, Oak Ridge, TN, for Department of Energy, U.S.

Olie, K., Vermeulen, P. & Hutzinger, O. (1977) Chlorodibenzo-p-dioxins and chlorodibenzo furans are trace components of fly ash and flue gas of some municipal incinerators in The Netherlands. *Chemosphere*, 6, 455–459.

Papageorgiou, A., Barton, J.R. & Karagiannidis, A. (2009) Assessment of the greenhouse effect impact of technologies used for energy recovery from municipal waste: A case for England. *Journal of Environmental Management*, 90, 2999–3012.

Patel, M.N. & Wheeler, P. (1994) *Fluidized Bed Combustion of MSW*. A status report for IEA, Task XI: MSW Conversion to Energy. Activity: MSW and RDF. Harwell, UK, ETSU.

Quina, M.J., Bordado, J.C.M. & Quinta-Ferreira, R.M. (2008a) Treatment and use of air pollution control residues from Municipal Solid Waste incineration: An overview. *Waste Management*, 28, 2097–2121.

Quina, M.J., Santos, R.C., Bordado, J.C.M. & Quinta-Ferreira, R.M. (2008b) Characterisation of air pollution control residues produced in a municipal solid waste incinerator in Portugal. *Journal of Hazardous Materials*, 152, 853–869.

Quina, M.J., Bordado, J.C.M. & Quinta-Ferreira, R.M. (2011) Air pollution control in municipal solid waste incinerators. In: Khallaf, M. (ed.) *The Impact of Air Pollution on Health, Economy, Environment and Agricultural Sources*. InTech. ISBN: 978-953-307-528-0. Available from: http://www.intechopen.com/books/the-impact-of-air-pollution-on-health-economy-environment-andagricultural-sources/air-pollution-control-in-municipal-solid-waste-incinerators.

Rademarkers, P., Hesseling, W. & Wetering, J. (2002) *Review on Corrosion in Waste Incinerators, and Possible Effect of Bromine*. TNO Industrial Technology.

Rhyner, C.R., Schwartz, L.J., Wenger, R.B. & Kohrell, M.G. (1995) *Waste Management and Resource Recovery*. Boca Raton, FL, CRC Press. ISBN: 0873715721, 9780873715720. 524 pp.

Rising, B. & Allen, J. (2002, revised 2004) *Emissions Assessment for RDF Combustion*. Washington, DC, U.S. Environmental Protection Agency, EPA/600/2-85/116.

Shepherd, P. & Gupta, B. (October 1992) *Data Summary of Municipal Solid Waste Management Alternatives, Volume IV: Appendix B—RDF Technologies*. Boulder, CO, NREL. Available from: www.nrel.gov/docs/legosti/old/4988d.pdf.

Solomon, N.G. (1998) Erosion-resistant coatings for fluidized bed boilers. *Materials Performance*, 37 (2), 38–43.

Sora, M.J. (January 7, 2013) *Incineration Overcapacity and Waste shipping in Europe: The End of the Proximity Principle*. Commissioned by Global Alliance for Incinerator Alternatives. Ventosa, I.P. (ed.).

Sorell, G. (1997) The role of chlorine in high temperature corrosion in waste-to-energy plants. *Materials at High Temperatures*, 14 (3), 137–150.

Spiegel, W. (2002) *Praxisnahe Unterstutzung der Betreiber bei der Optimierung und Nachhaltigen Nutzung der Bayerischen MV-Anlagen*. Augsburg, Germany, CheMin GmbH.

Stantec (March 2011) *Waste to Energy: A Technical Review of Municipal Solid Waste Thermal Treatment Practices*—Final report.

Takatsudo, Y., Nakamura, N., Ono, H., Mitsuhashi, M. & Kira, M. (1999) Advanced automatic combustion control system for refuse incineration plant. *MHI Technique*, 39 (2), 1999-5 (in Japanese).

Themelis, N.J. (2005) *Chlorine Balance in a Waste-to-Energy Facility*. New York, NY, Earth Engineering Center, Columbia University.

Thome-Kozmiensky, K.J. (1994) *ThermischeAbfallbehandlung*. EF-Verlag fuer Energie-und Umwelttechnik GmbH.

Thomsen, S.N. & Lundtorp, K. (Babcock & Wilcox Vølund A/S, Denmark) (2008) Optimization of a MSW incinerator. In: *2nd Intl. Symp. on Energy from Biomass and Waste, 17–20, Nov 2008, Fondazione Cini, Venice, Italy*. Available from: http://www.volund.dk/Waste_to_Energy/~/media/Downloads/Conference%20papers%20%20WTE/Optimization%20of%20a%20municipal%20solid%20waste%20incinerator.ashx.

U.S. EPA (June 1997) EPA 530-R-97-015 *Characterization of Municipal Solid Waste in the United States*.

U.S. EPA (October 2003) *Municipal Solid Waste in the United States: 2001 Facts and Figures*. Office of Solid Waste and Emergency Response, EPA530-R-03-011.

Uusitalo, M.A., Vuoristo, P.M.J. & Mantyla, T.A. (2002) Elevated temperature erosion-corrosion of thermal sprayed coatings in chlorine containing environments. *Wear*, 252 (7–8), 586–594.

Van Berlo, M. (2010) *An Example of Energy Efficiency—Amsterdam*. WtERT, Brno, Czech. Annual Meeting Europe.

Van Kessel, L.B.M. (2003) *Stochastic Disturbances and Dynamics of Thermal Processes, with Application to Municipal Solid Waste Combustion*. Doctoral Thesis. Eindhoven, The Netherlands, Eindhoven University of Technology.

Vehlow, J. (2015) Air pollution control systems in WtE units: An overview. *Waste Management*, 37, 58–74.

Waste-to-Energy Technology Council (WtERT) *Waste Incineration Plant*. Available from: http://www.wtert.eu/default.asp?Menue=12&ShowDok=13.

Wilson, A., Forsberg, U. & Noble, J. (1997) Experience of composite tubes in municipal waste incinerators. *Corrosion*, 97, 153.

World Bank (1999) *Decision Maker's Guide to Municipal Solid Waste Incineration*. Washington, DC, The International Bank for Reconstruction and Development/The World Bank.

World Bank Report (1999) *Municipal Solid Waste Incineration*. Washington, DC, Technical Guidance Report of World Bank.

WSP (2013) *Review of State-of-the-Art Waste-to-Energy Technologies, Stage 2—Case Studies*. Prepared by Kevin Whiting, WSP Environmental Ltd, WSP House, London, UK.

Yokoyama, T., Suzuki, Y., Akiyama, H., Ishizeki, K., Iwasaki, T., Noto, T., et al. (2001) Improvements and recent technology for fluidized bed waste incinerators. *NKK Technical Review*, 85, 38–43.

Zadeh, L.A. (1965a) Fuzzy sets. *Information and Control*, 8, 338–353.

Zadeh, L.A. (1965b) Fuzzy sets and systems. In: *Proc. Symp. System Theory, Polytechnic Inst. of Brooklyn, New York*. pp. 29–37.

Zhang, D.Q., Tan, S.K. & Gersberg, R.M. (2010) MSW management in China: Status, problems and challenges. *Journal of Environmental Management*, 91 (8), 1023–1033.

Zipser, S., Gommlich, A., Yang, Y.-B., Goh, Y.R., Zakaria, R., Nasserzadeh, V. & Swithenbank, J. (2002) Mathematical modeling of MSW Incinerator on a travelling bed. *Waste Management*, 22 (4), 369–380.

Zwahr, Y. (2003) Ways to improve the efficiency of waste to energy plants for the production of electricity, heat and reusable materials. In: *Proceedings of the 11th North America Waste to Energy Conference, Florida*.

Chapter 3

Pollutants and residues from thermal treatment

3.1 INTRODUCTION

The uncontrolled emissions generated from the combustion of hazardous and municipal solid wastes vary in composition subject to the combustor design and operating conditions and the waste composition. These emissions are of potential concern as they contain:

- Particulate matter (PM).
- Polychlorinated dibenzo-p-dioxins (PCDDs) and polychlorinated dibenzofurans (PCDFs), known collectively as dioxins and furans.
- Organic compounds (volatile, semivolatile, and nonvolatile) other than dioxins and furans, often referred to as products of incomplete combustion (PICs).
- Metals that are potentially toxic (e.g., mercury, lead, cadmium).
- Acid gases (e.g., hydrogen halides, nitrogen and sulfur oxides).

Table 3.1 provides an overview of the known human health effects of pollutants in each of these categories (U.S. EPA 2006).

Human and ecological exposure to emissions of these pollutants released to the atmosphere can occur through several means, for example, inhalation. The pathways of primary concern can vary by the pollutant that include PM, HCl, and CO. Other pollutants generated by combustion, such as NO_x, can undergo chemical reactions in the atmosphere with other compounds to form secondary pollutants, such as tropospheric ozone, for which inhalation exposure is a concern.

Other pollutants such as dioxins and furans, polycyclic aromatic hydrocarbons (PAHs), polychlorinated biphenyls (PCBs), and toxic metals are persistent since they do not quickly or ever degrade in the environment and have a tendency to bioaccumulate.

Human and ecological exposure to these pollutants can occur through ingestion of soil or biota that have been impacted by emissions of these pollutants. For example, dioxin compounds released to the air could deposit to the surface of a plant, this plant could be consumed by a cow or goat/lamb, which in turn is consumed by humans, resulting in exposure to the released dioxin compounds.

Besides, these emissions from incinerators and other industrial combustion processes are one of the primary sources of endocrine disrupting compounds (EDCs) in the environment, and evidence suggests that exposure to these EDCs may result in disorder of endocrine systems in both human and wildlife populations.

Table 3.1 Known health effects of constituents of combustion pollutant emissions.

Pollution category	Health effects
Dioxins/Furans	Short-term exposure to high levels may result in skin lesions and altered liver function. Chronic exposure may impair immune system, the developing nervous system, the endocrine system and reproductive functions; Identified as probable endocrine disrupting compounds (EDCs); Classified by EPA as a 'known human carcinogen' (WHO 1999; EPA 2004c); the contribution of combustors to EDCs in the environment is not known
Other PICs	Varies widely across chemicals. EPA has classified some PICs as 'probable human carcinogens' or 'known human carcinogens' and they are linked to a range of non-cancer effects such as impairment of the immune system and altered liver functions (EPA 2005d)
Metals	Linked to both chronic and acute health effects, although vary widely across chemicals. Mercury, if present, is linked to birth defects, immune system damage and nervous system disorders; lead is linked to nervous system disorders, and cadmium to kidney failure, hypertension and genetic damage
Acid gases	Linked to acute and chronic respiratory effects (EA 2005d)
Particulate matter (PM)	Linked to acute and chronic cardiopulmonary effects including premature mortality. Fine PM in combination with other pollutants are linked with a series of significant health problems that include premature death, respiratory related hospital admissions, aggravated asthma, chronic bronchitis and acute respiratory symptoms (EPA 1997a)
Polychlorinated Biphenyls (PCBs)	Shown to cause cancer, effect on immune system, reproductive system, endocrine system and other effects in animals; studies in humans provide supportive evidene for potential carcinogenic and non-carcinogenic effects (EPA 2004a)

Due to the persistence of some suspected EDCs in the environment and the consequent prospect of serious health effects in humans and wildlife at a global scale, the U.S. EPA offers a high priority to research on EDCs (EPA 2005c).

There are other residues formed in the form of bottom ash, grate siftings, boiler and economizer ash, fly ash and APCs during combustion process.

(a) *Bottom ash (also called slag)* consisting mainly of coarse non-combustible materials and unburned organic matter collected at the outlet of the combustion chamber in a quenching/cooling tank;

(b) *Grate siftings*, including relatively fine materials passing through the grate and collected at the bottom of the combustion chamber. Grate siftings are usually combined with bottom ash, as in most cases it is not possible to separate the two waste streams. Together they constitute about 20–30% mass of the original waste on a wet basis.

(c) *Boiler and economizer ash*, which represent the coarse fraction of the particulate carried over by the flue gases from the combustion chamber and collected at the heat recovery section. This residue may constitute up to 10% by mass of the original waste on a wet basis.

(d)   *Fly ash*, the fine particulate matter still in the flue gases downstream of the heat recovery units, is removed before any further treatment of the gaseous effluents. The quantity produced is around 1–3% of the waste input mass on a wet basis.

(e)   *Air pollution control (APC) residues*, including the particulate material captured after reagent injection in the acid gas treatment units prior to effluent gas discharge into the atmosphere. This residue may be in a solid, liquid or sludge form, depending on whether dry, semi-dry or wet processes are adopted for air pollution control. APC residues are usually in the range of 2% to 5% of the original waste on a wet basis (Sabbas et al. 2003, Chandler et al. 1997).

Due to the occurrence of volatilization and subsequent condensation as well as concentration during combustion, fly ash and APC residues contain high concentrations of heavy metals, salts as well as organic micro-pollutants.

Part of these residues (fly ash etc.) are sometimes used or recycled, but in general, they are sent to landfills. These landfills are exposed to precipitation which may dissolve soluble materials and result in leaching; and the leaching properties of a few materials may cause environmental problems. Hence, these residues and ash have to be treated before disposal in landfills. Currently bottom ash and fly ash are put to beneficial use after treatment, explained briefly in the later pages.

3.2   FORMATION OF POLLUTANTS

The formation of the main pollutants – particulate matter, heavy metals, acid gases (HCl, HF, HBr, and SO_2), CO, NO_x, and toxic organics, mainly dioxins/furans – during the combustion process are briefly explained.

Particulate Matter (PM) – The formation of PM (size and quantity) in MSW incineration depends on several factors: mainly, the waste characteristics, the method of waste feeding, primary air flow velocity, temperature, flue gas mixing, flue gas residence time, the incinerator design (grate type) and operation, and flue gas velocity. High combustion temperatures and long residence times allow for more complete burning of organic particles with a proportionate decrease in particle size. The boiler configuration and decreased gas velocity allow the larger, heavier particles to drop of the flue gases as they pass the boiler, but the majority move downstream (Sec. 4. Emission Estimates).

Actually, PM may originate from three sources: inorganic substances, organo-metallic substances, and unburned waste. Majority of the non-combustible inorganic materials exit the system primarily as bottom ash, with a small portion entrained in the flue gas stream and carried through the APC system. Organo-metallic compounds oxidize at high temperature and appear as inorganic oxides or metal salts in the flue gas. Unburned solid waste entrained in the flue gas is a consequence of incomplete combustion and agglomeration of small particles. Trace metals in the solid waste also become entrained in the flue gas, and a few trace metals may be found adsorbed onto entrained solids. At higher temperatures, with proper quantity of oxygen, the combustion of the fuel is more complete resulting in decreased particulate mass.

Facilities that operate with high primary air/secondary air flow ratios or relatively high excess air levels may entrain greater quantities of PM resulting in high levels.

RDF units carry higher PM from the furnace due to the suspension-feeding of the fuel.

Heavy Metals – Various metals are generally present in MSW streams. The metals such as As, Cd, Cr, Ni, Co, Cu, Mn and Pb, are emitted from the combustors in association with particulate matter usually as metal oxides and chlorides; and metals such as Hg as vapors. Cd in the waste arises from the discarded batteries, electronics and such products. The origin of Hg is attributed to many waste materials, electrical and electronic devices, thermometers, and so on. Mercury is by far the most thermally mobile metal, highly toxic, and at 357°C it is all volatilized into the flue gas. The primary measure of sorting out Hg-containing materials/objects from the waste feed as much as possible is the best way to avoid Hg vapour in the residues and flue gas.

A fraction of these metals can also be found in bottom ash, fly ash and sorbent. The fraction of each metal found in the particulate entrained in the flue gas versus that found in the bottom ash, is usually reflective of the volatility of the metal.

Due to the compositional variations in MSW feed stock, the metal concentrations also vary highly and are essentially independent of combustor type. The level of carbon in the fly ash appears to affect the level of Hg control. A high level of carbon in the fly ash can enhance Hg adsorption onto particles removed by the PM control device (USEPA website 10/96).

The Spray Drying Absorption (SDA), used to control trace metals in the flue gas, will condition the flue gas and reduce its temperature to 285–325°F. At that temperature, volatilized metals will condense on particulates to be collected in the fabric filter system.

Acid Gases – HCl and SO_2 are the main acid gases that form during the combustion of MSW. HF, HBr, and SO_3 are also present though in much lower concentrations. Concentrations of HCl and SO_2 in flue gases directly relate to the chlorine and sulfur content in the waste, but are considered independent of combustion conditions. Also, emissions of SO_2 and HCl in flue gases depend on the chemical form of sulfur and chlorine in the waste. The type of sulfur compounds formed and released from the furnace is dependent on the presence of other gaseous compounds, combustion temperatures, and the furnace's oxidizing or reducing conditions. In a typical of a solid waste combustor, excess oxygen generally allow the formation of SO_2 and SO_3 while oxygen deficient conditions result in hydrogen sulfide, carbonyl sulfide (COS) and elemental sulfur.

The major sources of chlorine in MSW are paper, plastics such as PVC, and other organic chlorides; and that of fluorine are plastics such as PTFE, fluorinated textiles, so on. The main sources of sulfur are the sulfur compounds in the waste such as asphalt shingles, gypsum wallboard, and tires. The chlorine and sulfur content vary considerably based on seasonal and local waste variations.

Sulfur dioxide, in the environment, may react with water vapor forming sulfuric acid which may further react to form sulfate salts, a particulate aerosol. Similarly, emitted chlorine reacts with water vapor to form hydrochloric acid. The presence of these acid gases in the atmosphere results in reduced visibility, material corrosion, sensitive organ irritation in humans and animals, and can cause acid rain and/or fog problems.

RDF processing does not generally impact the distribution of combustible materials in the waste fuel; as such, HCl and SO_2 concentrations for mass burn and RDF units are similar.

Carbon Monoxide – Carbon monoxide emissions result when all of the carbon in the waste is not oxidized to carbon dioxide. The oxygen concentration, the primary/secondary air flow ratio, furnace temperature, residence time and mixing (turbulence) are the factors that influence formation of CO. As waste burns in a fuel bed, it releases CO, hydrogen, and unburned hydrocarbons. Additional air (oxygen) then reacts with the gases escaping from the fuel bed to convert CO to CO_2 and H_2 to H_2O. If excess air is added to the combustion zone the local gas temperature is lowered retarding the oxidation reactions. If too little air is added, the mixing may be incomplete, allowing greater quantities of unburned hydrocarbons to escape the furnace. Both of the conditions would increase the CO emissions.

Because O_2 levels and air distributions vary among combustor types, CO levels also vary among combustor types. For example, semi-suspension-fired RDF units generally have higher CO levels than mass burn units, due to the effects of carryover of incompletely combusted materials into low temperature portions of the combustor, and, in some cases, due to instabilities that result from fuel feed characteristics.

CO concentration is a good indicator of combustion efficiency, and is a vital criterion for indicating instabilities and non-uniformities in the combustion process. It is during unstable combustion conditions that more carbonaceous material is available and higher dioxins/furans, and organic air pollutants occur. The high levels of CO corresponding to poor combustion conditions frequently correlate with high PCDD/PCDF emissions. When CO levels are low, however, correlations between CO and PCDD/PCDF are not well defined (due to the fact that many mechanisms may contribute to PCDD/PCDF formation), but their emissions are generally lower.

Nitrogen Oxides – Nitrogen oxides, NO_x, are formed when MSW containing nitrogen and its compounds is burned; the nitrogen is oxidized in the process. Nitric oxide (NO) is the primary component of NO_x. As in CO formation, NO_x formation can be greatly influenced by oxygen content in the air, furnace temperature, residence time and mixing in the combustor; and higher these values, higher the concentration of NO_x. But the CO formation is decreased. The NO_x formation can occur significantly for temperatures above $1300°C$ and whenever oxygen is not a limiting reagent. In the incineration plants, thermal NO_x is often much greater than fuel NO_x. As selective non-catalytic reduction (SNCR) uses ammonia as a reagent to reduce NO_x emissions in the post combustion flue gas stream, the SNCR/FGR system should be so designed and operated that ammonia escape is minimum.

Toxic Organic Compounds – A variety of organic compounds, including dioxins/furans, chlorobenzene (CB), PCBs, chlorophenols (CPs), and PAHs are present in MSW or can be formed during the combustion and post-combustion processes. These organic compounds are highly toxic, bioaccumulative and persistent in the environment. Organics in the flue gas can exist in the vapor phase or can be condensed or absorbed on fine particulates.

Dioxins/furans are formed in many processes including waste incineration. It was Olie and co-workers who showed that PCDD/PCDF were in trace quantities in fly ashes and flue gases of MSW incinerators (Olie et al. 1977). Since then, it is well known that dioxins are formed in trace quantities in incineration processes

(UNEP Chemical, 2005). Recently, Hites (2011), in an overview on dioxins, referred that PCDD/PCDF is established as a well known environmental contaminant family. McKay (2002) reviewed methods to minimise the formation of these very toxic compounds in municipal solid waste incineration.

PCDD/PCDF emissions from combustion sources can be explained by three principal mechanisms, not regarded as mutually exclusive (e.g., McKay 2002, AJ Chandler & Assoc. Ltd. 2006). In the first mechanism, PCDD/PCDF present as contaminants in the combusted organic material pass through the furnace and emit unaltered. The second mechanism, the *de novo* synthesis, considered the major mechanism, suggests that these pollutants are formed in the presence of fly ash containing chemically unrelated unburnt aromatics and metal catalysts. The reactions occur in the presence of oxygen and catalysts at temperatures 250°C to 450°C. For many incinerators, this temperature range is found only in the post furnace region, typically the waste heat boiler, or in electrostatic precipitators (ESP). *De novo* synthesis experiments suggested that more furan than dioxin congeners is formed.

In the third mechanism (precursor theory), these pollutants ultimately form from the thermal breakdown and molecular rearrangement of precursor ring compounds (chlorinated aromatic hydrocarbons) that have a structural resemblance to the PCDD/PCDF molecules. These ringed precursors that come from the combustion zone are products of incomplete combustion. The optimal temperature range for such formation is the same as that observed for *de novo* synthesis and also occur in the post-combustion zones of thermal processes. Therefore, the formation of PCDD/PCDF in thermal processes is undoubtedly the result of a complex set of competing chemical reactions.

Although chlorine is considered responsible for the formation of PCDD/PCDF in combustion systems, the empirical evidence indicates that for commercial incinerators, chlorine levels in the feed are not the dominant controlling factor for emission rates of PCDD/PCDF from the stack. There are complexities related to the combustion process itself, and some types of APC system tend to conceal any direct association. Therefore, the chlorine content of feeds is not a good indicator of PCDD/PCDF levels emitted.

Several studies have successfully correlated combustion process conditions to dioxin and furan formations (e.g., Fangmark et al. 1993, Gullett and Raghunathan 1997, Gullett et al. 1998, Wikstrom et al. 1999). Incomplete combustion is found to strongly promote the formation of chloroaromatic compounds (Blumenstock et al. 2000, Zimmermann et al. 2001, Aurell 2008), and many studies have established a positive correlation between PCDD/PCDF formation and levels of CO formed consequent to incomplete combustion (Bergstrom and Warman 1987; Kilgroe et al. 1990). It is to be noted that the average CO concentration seems to be of less significance than the occurrence and frequency of CO-peaks (Bergstrom and Warman 1987). It is also observed that single or a few CO-peaks though limited in height and duration, exert a sizeable influence on PCDD/PCDF levels and create favorable formation conditions for an extended time period (Wikstrom et al. 1999, Zimmermann et al. 2001, and Aurell et al. 2008). This may probably be related to soot deposits resulting in the incomplete combustion (Blumenstock et al. 2000). Fluctuations in the combustion process (CO peaks), high chlorine levels in the waste (1.7%) and low temperatures in the secondary combustion zone (660°C) all tended to increase the PCDD/PCDF emission levels. Also, the PCDD/PCDF ratio in the flue gas was found to depend on the $SO_2 - HCl$ ratio in the flue gas. The quench time profiles in the post-combustion zone and addition of sulfur were found to influence the formation pathways. Increased levels of chlorine in the waste improved the extent of chlorination of both dioxins and furans (Aurell 2008).

The dioxin formation can be effectively minimized to limit their emission concentrations well below 0.1 ng I-TEQ/m^3 by applying the following process conditions (McKay 2002): (a) combustion temperature: >1000°C; (b) chamber turbulence: Re > 50,000; (c) combustion residence time: >2 seconds; (d) post-combustion temperature – rapid quench cooling from 450 to 200°C; (e) using semi-dry scrubber, bag-filter, and activated carbon injection for air pollution control; (f) automated PCC system with interlocks and automatic MSW feed shut-off, for process control.

The reader may find extensive publications on the research results of dioxins and furans, especially their formation mechanisms and emissions from MSW incinerators, adverse effects on the environment, and their control (e.g., Hites 2011, AEA Technology 2012, Hartenstein and Licata 2008, AJ Chandler Assoc. 2006, Lopes et al. 2015).

3.3 FLUE GAS CLEANING/CONTROL SYSTEMS

(Vehlow 2015, WtERT 2014, WSP 2013, WB 1999a, Quina et al. 2011).

All waste-to-energy plants, based on combustion or other thermal processes, need an efficient gas cleaning for compliance with legislative air emission standards. The development of gas cleaning technologies started along with environment protection regulations in the late 1960s. Modern air pollution control (APC) systems comprise multiple stages for the removal of fly ashes, inorganic and organic gases, heavy metals, and dioxins from the flue gas. The main devices used for reduction of these pollutants are briefly described here. The history of the development of gas cleaning in waste-to-energy plants and how the current state-of-the-art APC systems have evolved are clearly explained in his recent paper by Vehlow (2015).

The main air pollutants formed in MSW incinerators that are under regulatory observation are: particulate matter, CO, TOC, HCl, HF, SO$_2$, NO$_x$, heavy metals, and dioxins/furans. Table 3.2 shows range of clean gas operation emissions levels reported from some European MSW incineration plants (EC 2006).

Environmental standards, including standards for emissions to the atmosphere differ from country to country. If standards are based on air quality considerations only, the air pollution problems are solved primarily by building tall stacks. Many countries fix emission standards that are considered technically and economically possible in that country; or considered state-of-the-art within emission control technology.

Primary and secondary measures can help reduce emission of pollutants. Primary measures actually obstruct the formation of pollutants, especially NO$_x$ and organics such as dioxins. These primary measures comprising an efficient combustion process (long flue gas retention time at high temperature with an appropriate oxygen content, intensive mixing and recirculation of flue gases, and so on), pre-precipitation of ashes in the boiler, and short flue gas retention time at intermediate temperatures must be applied. The content of CO and TOC ('total organic carbon' excluding CO) in the raw flue gas before inlet to the cleaning system is a good indicator of the efficiency of the combustion process.

Table 3.2 Range of clean gas operation emissions levels reported from some European MSW incineration plants.

Parameter	Type of Measurement C: continuous N: non-cont.	Daily averages (where continuous measurement used) in mg/m³		Half hour averages (where continuous measurement used) in mg/m³		Annual averages (mg/m³)
		Limits in 2000/76/EC	Range of values	Limits in 2000/76/EC	Range of values	Rauge of values
Dust	C	10	0.1–10	20	<0.05–15	0.1–4
HCl	C	10	0.1–10	60	<0.1–80	0.1–6
HF	C/N	1	0.1–1	4	<0.02–1	0.01–0.1
SO₂	C	50	0.5–50	200	0.1–250	0.2–20
NOₓ	C	200	30–200	400	20–450	20–180
NH₃	C	n/a	<0.1–3		0.55–3.55	
N₂O		n/a				
VOC (as TOC)	C	10	0.1–10	20	0.1–25	0.1–5
CO	C	50	1–100	100	1–150	2–45
Hg	C/N	0.05	0.0005–0.05	n/a	0.0014–0.036	0.0002–0.05
Cd	N	n/a	0.0003–0.003	n/a		
As	N	n/a	<0.0001–0.001	n/a		
Pb	N	n/a	<0.002–0.044	n/a		
Cr	N	n/a	0.0004–0.002	n/a		
Co	N	n/a	<0.002	n/a		
Ni	N	n/a	0.0003–0.002	n/a		
Cd and Tl	N	0.05		n/a		0.0002–0.03
∑ other metals 1	N	0.5		n/a		0.0002–0.05
∑ other metals 2	N	n/a	0.01–0.1	n/a		
Benz(a)pyrene	N	n/a		n/a		<0.001
∑ PCB	N	n/a		n/a		<0.005
∑ PAH	N	n/a		n/a		<0.01
PCDD/F (ng TEQ/m³)	N	0.1 (ng TEQ/m³)		n/a		0.0002–0.08 (ng TEQ/m³)

[1] In some cases there are no emission limit values in force for NOₓ. For such installations a typical range of values is 250–550 mg/Nm³ (discontinuous measurement).
[2] Other metals 1 = Sb, As, Pb, Cr, Co, Cu, Mn, Ni, V
[3] Other metals 2 = Sb, Pb, Cr, Cu, Mn, V, Co, Ni, Se and Te
[4] Where non-continuous measurements are indicated (N) the averaging period does not apply. Sampling periods are generally in the order of 4–8 hours for such measurements.
[5] Data is standardised at 11% Oxygen, dry gas, 273K and 101.3 kPa.

Table 3.3 Pollutants and related removal systems.

Pollutant	Removal systems
Particulates	Cyclones and multi-cyclones, Fabric filters (FF), Electrostatic Precipitators (ESP), Wet ESPs & wet scrubbers
Nitrogen Oxides (NO_x)	Primary measures: Air & temperature control; Flue gas recirculation, Secondary measures: SNCR, SCR
Acid gases (SO_2, HCl, HF etc.)	Wet scrubber, Semi-dry scrubber + FF, Dry scrubbers, Fabric filters; SDA
Heavy metals	Fabric filters, Activated carbon injection; all gadgets used for particulates
Toxic organics (mainly Dioxins & Furans)	Flue gas recirculation, fabric filters, Activated carbon injection, SDA
Mercury (Hg)	Primary measure: sorting out Hg- contained substances from the waste; Secondary measures: scrubbers by adding oxidants, activated carbon, furnace coke or zeolite

The secondary measures are the flue gas treatment systems (also called Air Pollution Control system, APC), consisting of Electrostatic precipitators; Fabric (bag house) filters; dry, semidry, and wet scrubbers (acid gas removal systems); catalysts; or multiple units/stages, designed for specific emission removal. A few characteristics of the secondary measures are that they precipitate, adsorb, absorb, or transform the pollutants.

Combinations of several individual cleaning stages shown in Table 3.3 for removal of different pollutants, provide an overall treatment system, with the actual number of potential flue gas treatment combinations reaching the 408 (European Commission, 2006).

The selection of the APC system depends primarily on the actual emission limits or standards, and the desired emission level. 'Basic' emission control involving reduction of the particulate matter only is a minimum requirement and simple to operate and maintain with low investment. Moving from basic to 'medium' or 'advanced' emission control allow an increased efficiency; but it needs to be evaluated in the light of increased complexity, the quantity and types of residues, and the capital and operating cost. The state-of-the-art flue gas cleaning systems utilized especially in Europe and the U.S. are quite complex complying with the country's emission limits/standards and providing higher benefits (WB 1999a, EC 2006). The air emission limits as applied in Europe are given elsewhere (WSP 2013).

It is common practice to arrange the APC systems by installing subsequent process stages to remove the pollutants in the following order: fly ash; acid gases; specific contaminants like Hg or PCDD/PCDF; and nitrogen oxides (IEABioenergyTask36, 2007–2009).

Flue Gas Recirculation (FGR)

FGR is a part of the furnace design, designed to lower the excess air rate, which increases thermal efficiency and reduces the size of the downstream gas clean-up equipment. After passing through the dust filter, about 20 to 30% of flue gas is limited and retained through an insulated duct to the furnace. The recirculated flue gas is injected through separate nozzles in the furnace and in the turbulence zone at the inlet to the secondary combustion chamber, the first pass of the boiler. FGR provides additional advantages such as (a) lesser formation of thermal NO_x as less nitrogen is delivered to the combustion chamber due to lower excess air rate, (b) reduced dioxin formation, and (c) improved flow and turbulence conditions.

Flue gas recirculation requires all duct connections be properly welded and maintained to avoid leaks which otherwise cause corrosion (WB 1999a).

3.3.1   Particle (PM) removal

The first step in the flue gas treatment is to remove the solid particles of sizes, 1 μm to 1 mm applying the common practice of using cyclones, fabric filters or electrostatic precipitators (ESP).

Inside the cyclone, the gas entering a cylindrical chamber swirls around an immersed tube and due to inertia, the particulates are carried to the cylinder wall. From there they exit through the conical section on the bottom while the clean gas exits through the tube (Figure 3.1) (Bilitewski et al. 1997).

Electrostatic precipitators utilize high voltage to electrically charge the particles contained in the flue gas by hanging wires vertically in the gas stream; and the charged particles move toward the electrode forming a dust layer on it. The electrode is shaken to push the dust particals down. The ESP is robust and simple to operate. It is very effective in removing larger particulates but has limited efficacy at removing small particles.

Fabric filter contains a large number of filter bags. As the raw flue gas passes through the filter bags, the particle material is retained on the walls and the cleaned gas flows through and exit at the top. The particles remain until compressed air is blown in the opposite direction, cleaning the filter and causing the dust to fall down where it is collected.

Electrostatic precipitators and fabric filters have a similar operational area, with the latter showing a high particulate removal efficiency and a better performance for removing particles smaller than 1 μm (see Figure 3.2).

As the fabric filters also provide a surface for the reactions to occur to neutralize acid gases, they are usually located at the back of the APC system downstream of the scrubber. Due to these advantages, fabric filters are preferred in all modern waste incineration plants. But, in plants operating in areas with stringent emission limits, both fabric filter and ESP are installed. Cyclones, due to their limited removal efficiency, are not applied currently to modern waste incinerators. However, they are utilized as a complement with other flue gas treatment stages (TWG Comments 2003, Bilitewski et al. 1997).

3.3.2   Gaseous contaminants removal

The next stage is the removal of gaseous contaminants in the flue gas. All the current technologies are based on either absorption or adsorption processes. In the absorption process the flue gas is mixed with additives that react and transform the contaminant gases into nonpolluting products; in adsorption processes the molecules of the contaminant gases attach to the surface of another material allowing the cleaned air to

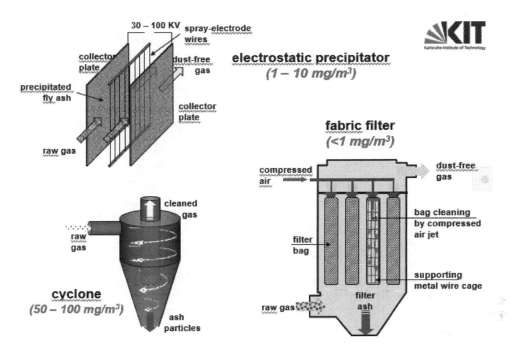

Figure 3.1 Schematics of three common solid particle removers: Cyclones, ESP, Fabric filter (source: Vehlow 2012; reproduced with the permission of Dr. Juergen Vehlow, ITC).

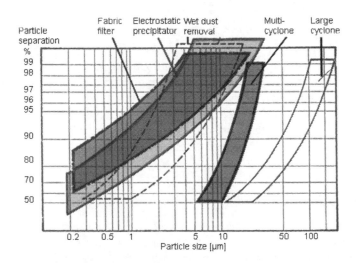

Figure 3.2 Operational Areas of Dust Removing Processes (source: WtERT 2014; reproduced with permission from Prof. Dr. Rudi Karpf, WtERT, ete.a Ingenieurgesellschaft für Energie und Umweltengineering & Beratung mbH).

flow. Scrubbers are used for the removal of the acidic and alkaline gaseous pollutants which can be wet type or semi-dry type or dry type.

(a) Wet scrubbers

In a wet scrubber, the flue gas is fed into water, hydrogen peroxide, and/or a washing solution containing part of the reagent (e.g. sodium hydroxide solution). The reaction product is aqueous.

The scrubber solution is strongly acidic (typically pH 0–1) due to acids formation in the process of deposition. HCl and HF are mainly removed in the first stage of the wet scrubber. The effluent from the first stage is recycled many times, with small fresh water addition and a bleed from the scrubber to maintain acid gas removal efficiency.

Removal of SO_2 is achieved in the washing stage controlled at a pH close to neutral or alkaline (generally pH 6–7) in which caustic soda solution or lime milk is added. This removal takes place in a separate washing stage, where further removal of HCl and HF also occurs.

The two-stage systems have very high removal efficiencies for HCl, HF and HBr, mercury, and SO_2. The raw gas concentrations of these components are easily reduced well below the emission standards. The solid scrubbing residues are removed from the gas flow in a subsequent fabric filter. Another way to evaporate the scrubbing solutions is by drying in steam heated devices.

The advantages are, low consumption of chemicals, use of inexpensive $CaCO_3$ to neutralize most of the HCl, and small quantity of solid residues; the disadvantages are waste water discharge, lining of quencher and scrubber with special materials to avoid risk of corrosion, and forming dense white cloud (if no reheating).

(b) Semi-dry scrubbers

Semi-dry (wet-dry) scrubbers, unlike wet scrubbers, do not saturate the flue gas stream being treated. In this type, an aqueous absorption agent sprayed into the flue gas reacts with the acid gases; and the water solution evaporates and the reaction products which are solids are filtered (Thomé-Kozmiensky 1985).

The advantages of this type are: no waste water, less prone to corrosion, and no visible plume; the disadvantages are expensive NOx removal process, high consumption of energy and chemicals, more solid residues, and the dioxins are only adsorbed, not destroyed.

(c) Dry scrubbers

The principle is more similar to semi-dry scrubbers. In dry sorption processes, the flue gas passes through a fine and dry powder, and the solid reaction products generated are removed from the flue gas stream by a filter as shown in Figure 3.3 (Bilitewski et al. 1997, UBA 2001). Calcium hydroxide or lime, and sodium bicarbonate are the typical absorption agents utilized. The ratio used with lime is typically two or three times the stoichiometric amount of the substance to be deposited, while with sodium bicarbonate the ratio is lower (European Commission 2006). The typical reactions of

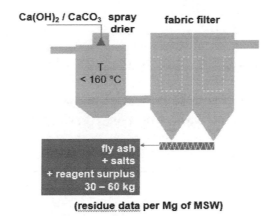

Figure 3.3 Schematic diagram of a semi-dry scrubbing system for FG cleaning (source: Vehlow 2012; reproduced with the permission of Dr. Juergen Vehlow, ITC).

lime with acidic components that occur in dry sorption:

$$Ca(OH)_2 + SO_2 \rightarrow CaSO_3 + H_2O$$

$$Ca(OH)_2 + SO_2 + \tfrac{1}{2}O_2 \rightarrow CaSO_4 + H_2O$$

$$Ca(OH)_2 + 2HCl \rightarrow CaCl_2 + 2H_2O$$

$$Ca(OH)_2 + 2HF \rightarrow CaF_2 + 2H_2O$$

The advantages are no waste water, less tendency to corrode, easily adjustable to meet the requirements of the advanced control level, and only visible plume in very cold weather. The disadvantages are higher consumption of chemicals, use of relatively expensive $Ca(OH)_2$, more solid residues, and in medium control level reduction in the removal of SO_2.

The emission limits of the advanced control level may be achieved by the dry and semi-dry systems, but with increased chemicals consumption. The advanced wet system consists of a combination of ESP, gas/gas heat exchanger, two-stage scrubbing, and fabric filter.

The treatment efficiencies of dry and semidry systems for removing HCl, HF, and SO_2 depend on the quantity of lime consumed. A completely dry system will, however, need lime in excessive quantities resulting in higher production of residues. The Hg and dioxin removal limits may be realized by adding activated carbon to the lime.

The advanced wet system has an additional wet scrubber in which SO_2 is reduced by reaction with NaOH solution or $CaCO_3$ suspension due to the excess oxygen in the flue gas. The reaction product in the former case is sodium sulfate (Na_2SO_4) solution, and in the later, gypsum ($CaSO_4$, $2H_2O$) suspension.

If NaOH is applied, the scrubber system must have an additional water treatment plant in which the sulfate ions of the Na_2SO_4 solution are precipitated as gypsum by Ca ions (for example, by mixing with $CaCl_2$ contained in the water treated for HCl removal). If $CaCO_3$ is used, the gypsum is formed directly and may be removed as sludge and dewatered.

The gas from the SO_2 scrubber is reheated in the gas heat exchanger and led to a fabric filter before the activated carbon or a mixture of lime and activated carbon is injected into the duct. If the gas now penetrates the filter, Hg and dioxins are removed to concentrations well below the advanced control level limits. In addition, the dust, HCl, HF, SO_2 and the other heavy metals are further reduced.

3.3.3 Heavy metal and dioxin/furan removal

Volatile heavy metals (Hg, Cd, Pb) and dioxin/furans remaining in the flue gas stream which are non-soluble in water are minimized by the injection of activated carbon. This material has an exceptionally high specific surface area and is very effective at adsorbing such compounds. Activated carbon is usually co-injected in a dry or semi-dry scrubber as already mentioned. Proper design and operation of the combustion chamber accomplish the control of organics. Good combustion practices (GCP) which consist of proper 3Ts (time, temperature and turbulence) in the combustion chamber and/or in combination with post-combustion flue gas controls would reduce these pollutants. GCP includes providing sufficient oxygen for the destruction of organic species, limiting PM carryover, and monitoring PM inlet temperature to minimize condensation of vapor phase constituents. The oxygen concentration should be low whereas temperature and residence time should be high for their minimization.

Prolonging the flue gas residence time at a relatively high temperature (460°C) reduce the dioxin/furans emissions most significantly. It was also found that by increasing the SO_2-to-HCl ratio to 1.6 in the flue gas could reduce the dioxin levels, but not the furan levels (Aurell 2008).

Using 4-year monitoring data of a waste incinerator in Taiwan, Bunsan et al. (2013) developed a simplified monitoring model to predict and to control dioxin emissions, based on an artificial neural network (ANN). The results indicated the prediction model based on a back-propagation neural network is a promising method to deal with complex and non-linear data with the help of statistics in screening out the useful variables for modeling.

3.3.4 Nitrogen Oxide Removal

The formation of thermal NO_x can be minimized by ensuring combustion control and reducing excess air for combustion. This requires (a) good mixing of waste so that the feed to the combustion chamber is as homogeneous as possible; (b) low excess oxygen (including use of FGR); and (c) stable, low temperature in the combustion chamber (an advantage of using a fluidised bed system) (WSP 2013).

Nitrogen oxides, once formed, can be destroyed either by selective non-catalytic reduction (SNCR) or by selective catalytic reduction (SCR). SNCR applies dry urea $[CO(NH_2)_2]$ or NH_3 as reducing agent directly in the furnace. At temperatures between 900 and 1050°C the reducing agents react with the nitrogen oxides to form water and nitrogen.

$$4NO + 4NH_3 \rightarrow 4N_2 + 6H_2O$$

$$6NO_2 + 8NH_3 \rightarrow 7N_2 + 12H_2O$$

$$CO(NH_2)_2 + H_2O \rightarrow 2NH_3 + CO_2$$

The high temperatures required for SNCR means ammonia must be injected into the upper part of the combustion chamber where the effectiveness of SNCR is limited by conditions in this region. The quantities of ammonia required to achieve higher reductions lead to carryover of unreacted ammonia into the flue gas stream resulting in undesirable emissions from the process.

However, it is possible to remove these emissions using a wet scrubber. In SCR (Jungtten et al. 1988), a catalyst such as mixture of ammonia and air reacts with the flue gas to form oxygen and water. These reactions take place at temperatures, 200–400°C which requires the flue gas to be reheated to at least 180°C before passing through the SCR unit and to the stack. Hence, there is a loss of efficiency with this type of system (WSP 2013). The SCR module must be installed after the units for removal of particulates and acidic gases. The advantages of SNCR are lower investment cost and less corrosion problems, and can achieve reduction rates upto 70% (typical 30 to 60%) while SCR is more efficient and can achieve up to 85% (typical 50 to 80%) (Bilitewski et al. 1997, Quina et al. 2011).

$$4NO + 4NH_3 + O_2 \rightarrow 4N_2 + 6H_2O$$

$$6NO + 4NH_3 \rightarrow 5N_2 + 6H_2O$$

$$2NO_2 + 8NH_3 + O_2 \rightarrow 3N_2 + 6H_2O$$

3.3.5 Spray Drying Absorption (SDA)

The SDA process is a semi-dry flue gas desulphurization process. The process uses slaked lime $Ca(OH)_2$ as absorbent and results in a stable and dry end product, mainly consisting of fly ash and various calcium compounds. SDA is a precise, effective system to remove multiple pollutants in one common process.

Hot, raw flue gas is fed into a 'spray drying absorption' chamber where it comes into contact with a fine spray of alkaline slurry (usually slaked lime). Virtually all the acidic components in the flue gas (HCl, HF, SO_2, and SO_3) are absorbed into the alkaline droplets, while the water is evaporated simultaneously. Precise control of the gas distribution, slurry flow rate and droplet size ensures droplets are converted into a fine powder. The injection of activated carbon into the flue gas can be used to enhance the removal of *mercury* and *dioxins*. Some fly ash and reaction products drop to the bottom of the absorber and are discharged. The treated flue gas continues on to a dust collector, where any remaining suspended solids are removed. The cleaned gases then exit through the stack.

The chemistry of SDA process can be represented as:

$$2HCl + Ca(OH)_2 \rightarrow CaCl_2 + 2H_2O$$

$$2HF + Ca(OH)_2 \rightarrow CaF_2 + 2H_2O$$

$$Ca(OH)_2 + SO_2 + H_2O \rightarrow CaSO_3 \cdot \tfrac{1}{2}H_2O + \tfrac{3}{2}H_2O$$

$$CaSO_3 \cdot \tfrac{1}{2}H_2O + \tfrac{1}{2}O_2 + \tfrac{3}{2}H_2O \rightarrow CaSO_4 \cdot 2H_2O$$

SDA is a self-adapting system to changes in flue gas flow rate, temperature and composition, and is suitable for incinerators of all types and sizes, requiring only a

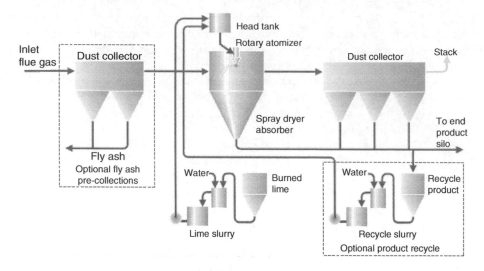

Figure 3.4 Schematic of Spray Drying Absorption process (source: GEA Niro Brochure).

single absorber and a single rotary atomizer per incineration line. In addition, the SDA process can be easily integrated with other flue gas cleaning technologies. The basic SDA process can be enhanced by additional process options in order to meet the specific needs of incinerators with a very high concentration of pollutants and/or very strict emissions requirements.

SDA is used in more than 160 incineration facilities worldwide. It is the most frequently used acid gas control technology for waste incinerators in the United States. When used in combination with an ESP or FF, the system can control dioxin/furans, particulates (and metals), SO_2, and HCl emissions. However, Spray dryer/fabric filter systems are more common than SDA/ESP systems and are used mostly on new, large MSW incinerators.

3.3.6 Dry sorbent injection

This technology has been developed primarily to control acid gas emissions. If combined with flue gas cooling and an ESP or FF, these processes may also control dioxins/furans and particulate emissions. Two primary types of dry sorbent injection technologies exist; and the more widely used one is 'duct sorbent injection (DSI)'. It involves injecting dry alkali sorbents into flue gas downstream of the combustor outlet and upstream of the PM control device. The second type called 'furnace sorbent injection (FSI)', injects sorbent directly into the combustor.

In DSI, powdered sorbent is pneumatically injected into either a separate reaction vessel or a section of flue gas duct located downstream of the combustor economizer or quench tower. Alkali in the sorbent (generally Ca or Na) reacts with HCl, HF, and SO_2 to form alkali salts (e.g., calcium chloride, calcium fluoride, and calcium sulfite). By lowering the acid content of the flue gas, downstream equipment can be operated at reduced temperatures thereby minimizing the potential for acid corrosion of the equipment. Solid reaction products, fly ash, and unreacted sorbent are collected with either ESP or FF. The acid gas removal efficiency with DSI depends on the method of sorbent injection, flue gas temperature, sorbent type and feed rate, and the extent of sorbent mixing with the flue gas. As all the DSI systems are not of the same design, performance of the systems varies. Flue gas temperature at the point of sorbent injection can range from about 150 to 320°C (300 to 600°F) depending on the sorbent used and the design of the process. Successfully tested sorbents include hydrated lime $(Ca(OH)_2)$, soda ash (Na_2CO_3), and sodium bicarbonate, $NaHCO_3$. With hydrated lime as sorbent, some DSI systems are observed to have achieved removal efficiencies comparable to SDA systems, although performance is generally lower.

By combining flue gas recirculation (FGR) with DSI it may be possible to increase dioxin/furans removal through a combination of vapor condensation and adsorption onto the sorbent surface. Cooling may also benefit PM control by decreasing the effective flue gas flow rate and reducing the resistivity of individual particles.

Furnace sorbent injection (FSI) involves the injection of powdered alkali sorbent (lime or limestone) into the furnace section of a combustor. This can be accomplished by addition of sorbent to the overfire (secondary) air or injection through separate ports or mixing with the waste prior to feeding to the combustor. As with DSI, either ESP or FF is used to collect the reaction products – fly ash and unreacted sorbent.

The basic chemistry of FSI is similar to DSI. Both use a reaction of sorbent with acid gases to form alkali salts. However, several key differences exist between them. First, by injecting sorbent directly into the furnace at temperatures of 870 to 1200°C, limestone can be calcined in the combustor to form more reactive lime, thereby allowing use of less expensive limestone as a sorbent. Second, at these temperatures, SO_2 and lime react in the combustor, thus providing a mechanism for effective removal of SO_2 at relatively low sorbent feed rates. Third, by injecting sorbent into the furnace rather than into a downstream duct, additional time is available for mixing and reaction between the sorbent and acid gases. Fourth, if a significant portion of the HCl is removed before the flue gas exits the combustor it may be possible to reduce the formation of dioxins/furans in later sections of the flue gas ducting. However, HCl and lime do not react with each other at temperatures above 760°C which is the flue gas temperature that exists in the convective sections of the combustor. Therefore, HCl removal may be lower with FSI compared to DSI. The most disadvantages of FSI include fouling and erosion of convective heat transfer surfaces by the injected sorbent.

3.3.7 Moisture condensation in gas cleaning systems

Moisture condensation is one of the major problems in gas cleaning systems. It causes severe corrosion problems as well as fly ash aggregation as mud, and leads to several gas cleaning steps. Careful design and operation of the incineration systems are essential to avoid moisture condensation. Otherwise, constant failures reduce the life span of the overall plant as well as the average availability due to more recurrent maintenance necessities.

The overall water balance for steady and dynamic conditions needs to be assessed in addition to carefully estimating the dew point of the flue gas in different critical points. The moisture in the flue gas comes from: (i) moisture in the inlet air, (ii) moisture vaporized from the wastes, (iii) water formed in the combustion reactions, and (iv) water vaporized in the flue gas stream in the wet or semi-wet cleaning steps. The waste at the collection points must be protected from rain or snow to decrease the effect from (i) and (ii). Absolute prevention of condensation is highly important at the design stage. Further, the probability of the condensation occurring becomes higher toward the last stages of the cleaning system as the flue gas temperature lowers as it travels from the outlet of the boiler to the stack, partly due to the energy recovering systems. In practice, re-heating of the flue gas is done to avoid condensation by one of these methods: (i) Injection of limited flow of hot flue gas that boosts the main stream, an effective system of re-heating, but for the problem of a slight decrease of overall cleaning efficiency as the hot flue gas from the boiler is more contaminated; or (ii) By heat exchanging with hot flue gas wherein the two streams are kept separated; the problem is the increased pressure drop throughout the heat exchanger and the higher investment cost. In both cases, rise of 2 to 5°C over the higher dew point temperature that can occur, is effective to prevent condensation since the pressure profile is, as a rule, very stable (Quina et al. 2011).

The re-heating is often considered in the design either (i) immediately before the exhaust stack to avoid the formation of plume caused by condensation arising from cooling by the cold air, or (ii) before the hose fabric filters to avoid clogging by mud formation in the inside the filter wall. The good design practice is to use corrosion resistant materials and special coatings where condensation is more prone to occur, but stopping the occurrence is certainly more significant.

3.4   NEW DEVELOPMENTS IN FLUE GAS CLEANING

3.4.1   Alstom's NID (Novel Integrated Desulfurization) system

It is a humidified dry system and uses unslaked lime and activated carbon as reagents. Unslaked lime is charged from a silo to the hydrator, where a small amount of water is sprayed on the lime, which partly hydrates. It overflows to the next vessel, a mixer, where further slaking takes place with three water spraying nozzles. Slaked lime, which is still freely fluent, flows into a vertical, rectangular flue gas duct, which is the reactor. This dust mixture is entrained with flue gas flow up to the fabric filter. During the 20 metre flow in the reactor duct and on the filter surface cake, chemicals react with the acidic gases and harmful pollutants in the flue gas. Dust mixture is mainly recirculated through the mixer to the reactor and partly discharged to the ash silo. The main acidic gases SO_2 and HCl form dry salts $CaSO_3$, $CaSO_4$ and $CaCl_2$. A part of the cleaned gas is recirculated to the reactor to increase the flue gas velocity, if needed. Thus it is possible to capture the acid gases (SO_2, HCl, HF and HBr). However, since NID is a dry process it is quite easy to add additional adsorbents to capture other harmful components. For this purpose activated carbon is used which captures heavy metals as well as dioxins/furans. The NID system is thus able to capture the acid gases, heavy metals, dioxins and the fine ash particles. The schematic representation of the process is shown in Figure 3.5.

Desulphurization:

$SO_2 + Ca(OH)_2 \rightarrow CaSO_3 \cdot \frac{1}{2}H_2O + \frac{1}{2}H_2O$ in humid conditions (liquid water or adsorbed humidity)

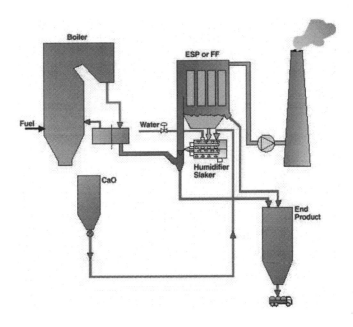

Figure 3.5 Schematic representation of Alstom's NID process (Source: Werther J. 2007; reproduced with the permission of Dr. Joachim Werther).

HCl removal: $2HCl + Ca(OH)_2 \rightarrow CaCl_2 \cdot 2H_2O$; and
Removal of heavy metals and dioxins/furans: By treating with activated carbon.

The main advantages of the system are (a) High SO_2 removal efficiency; (b) Lower operating cost than comparable systems; (c) Lower investment than ESP + Wet FGD; (d) No separate Dust Removal system necessary; (e) Less maintenance due to less equipment and no corrosion, (f) No water spraying and (g) 100% removal of SO_3.

3.4.2   Lurgi Lentjes Circoclean process

The Circoclean process developed by Lurgi Lentjes (Yi et al. 2005) is somewhat similar to Alstom's NID process. The main difference is the shape of the reactor: the flue gas enters through the bottom of the venturi shaped absorber which is operated as a circulating fluidized bed. The optimal reaction temperature which is 20–30°C above the wet bulb temperature is achieved by water injection directly into the bottom of the fluidized bed. For operation in part load of the boiler clean gas can be recirculated to ascertain a stable operation of the CFB. The flow diagram is shown in Figure 3.6.

3.4.3   Combined scrubbing process by Forschungszentrum Karlsruhe

A more sophisticated process has been developed in the Forschungszentrum Karlsruhe (Korell et al. 2006) for the combined removal of mercury, dioxins and particulate fines and aerosols. Figure 3.7 shows this multi-purpose system. After passing a quench, the

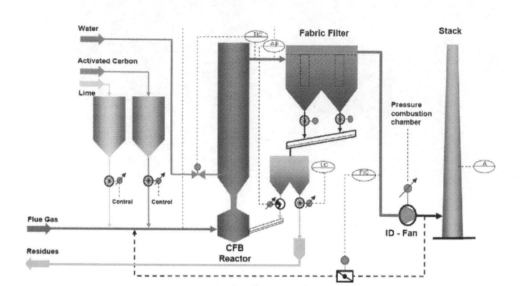

Figure 3.6 Schematic of Lentjes Circoclean process (source: Werther J. 2007; reproduced with the permission of Dr. Joachim Werther).

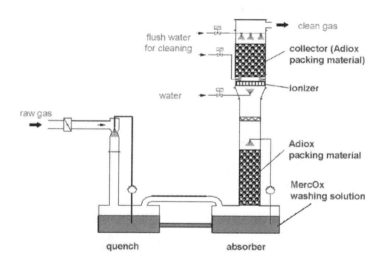

Figure 3.7 The combined scrubbing process developed by Forschungszentrum Karlsruhe (source: Werther 2007; reproduced with the permission of Dr. Joachim Werther).

flue gas enters a scrubber where it comes into contact with a H_2O_2 saturated washing solution which oxidizes the elemental mercury according to the MercOx® process.

The scrubber contains an Adiox® packing which causes the dioxins to be absorbed in a plastic material and adsorbed at carbon particles embedded in the plastic. From the scrubber, the flue gas passes through an ionizing section where fine particles and aerosols are electrically charged which subsequently precipitate on the surface of an Adiox® packing.

Excellent performance was reported when the process was tested in the bypass of Karlsruhe's THERESA incinerator. It is installed in the waste incinerator plant in Sweden and is operating successfully.

3.5   ENERGY & WATER CONSUMPTION

The water and energy requirements in a waste incinerator are briefly summarized (Waste-C-Control, at www.epem.gr/waste-c-control/database/html/WtE-01.htm).

(a) Energy

The incineration process itself requires energy for its operation at several stages. The demand varies to a great extent depending on the construction of the plant. The factors that increase the energy demand, in particular, are:

(a)   Mechanical pre-treatment, e.g. shredding and other waste processing methods
(b)   Incineration air preheating
(c)   Reheat of flue-gas (for gas treatment devices or plume suppression)
(d)   Operation of waste water evaporation plant
(e)   flue-gas treatment (FGT) systems with high pressure drops (e.g. filtration systems) which require higher-power forced draught fans
(f)   Decrease in the net heat value of the waste resulting in the necessity to add additional fuels to maintain the required minimum combustion temperatures, and
(g)   Sludge treatment (drying).

The WTE facilities in Europe (European Commission 2006b) consume energy ranging from 0.062 MWh to 0.257 MWh per tonne of waste feedstock for electricity; and 0.021 MWh to 0.935 MWh per tonne of waste incinerated. McDougal et al. (2002) reported a specific electricity consumption of 70 kWh and 0.23 Nm^3 of natural gas per tonne of waste combusted during start-up for incinerator heating up. The energy consumption of the Plant is also influenced by the lower heating value of the waste. This is largely due to increased flue-gas volumes with higher net calorific waste, which require larger FGT capacity.

There are significant local variations in the amount of power generation; typically 400 to 700 kWh of electricity can be generated with one tonne of waste in a MSW incineration plant. This is, of course, dependent upon the size of the plant, steam parameters, extent of steam utilization, and largely on the LHV of the waste.

The LHV of the feed waste to incinerators can be calculated from the composition using material specific LHV, available in the literature.

The amount of energy available for export usually depends upon the amount produced and the amount self-consumed by the Plant which can itself vary significantly. The FGT system consumption is often significant and varies with the type of system applied based on emission levels required. In some cases, the energy required to run the Plant is imported from external supply, with all of that generated by the Plant being exported to the grid – the local balance usually reflects local pricing for the electricity generated compared to general grid prices.

Regarding heat recovery IPPC BREF reports heat production ranging from 1.376 MWh to 2.511 MWh per tonne of waste incinerated, while heat exported in the grid ranges from 0.952 MWh to 1.786 MWh per tonne. In the case of combined electricity/heat generation, approx. 1250 kWh of additional heat per tonne of waste can be used at full load (EC 2006b).

(b) Water

Water is used in waste incineration plants for different purposes – flue gas treatment, steam production etc. – though the main consumption of water is for flue-gas cleaning. Dry systems consume the least water and wet systems generally the most. Semi-wet systems fall in between. Typical effluent rates are around 250 kg/t of waste treated (wet scrubbing, other flue gas cleaning technologies provide different figures). It is possible for wet systems to reduce consumption greatly by re-circulating treated effluent as a feed for scrubbing water. This can be performed to a certain degree only as salt can build up in the re-circulated water. The use of cooled condensing scrubbers provides another way of removing water from the flue-gas stream, which after treatment, can be re-circulated to the scrubbers. The salt build up remains an issue. Processes that use 'boilers with no energy recovery' may consume more water, up to 3.5 tonnes water per tonne of waste because the required cooling of flue-gas is carried out using water injection. Installations with a rapid quench system may use up to 20 tonnes of water per tonne of waste incinerated.

3.6   MANAGEMENT OF RESIDUES

The main products from the waste incineration are bottom ash, fly ash and APC residue. In most modern WTE facilities, fly ash is mixed together with APC residues, and are both collectively referred to as APC residues (Stantec 2011). Bottom ash is considered separately because of different composition and nature.

Several efforts have been made to improve the environmental quality of waste incineration residues and to recycle or utilise at least part of specific residue flows. Both in-process and post-treatment techniques are applied. In-process measures are aimed at changing the incineration parameters in order to improve burnout or to shift the metal distribution over the various residues. Post-treatment techniques include: ageing, mechanical treatment, washing, thermal treatment and stabilization (EC 2006, Vehlow 2002).

Lam et al. (2010) reviewed the characteristics of municipal solid waste incineration ashes and the possible treatment methods for their utilization. The incineration ash utilization in cement and concrete production, road pavement, glasses and ceramics, agriculture, stabilizing agent, adsorbents and zeolite production are also discussed.

3.6.1   Treatment & use of bottom ash

The bottom ash from the waste incineration plant is a mixed material that may contain varying proportions of glass, ceramics, metals, brick and concrete in addition to clinker and ash. The later can be usefully utilized. It is gravel-type, unlike other MSW inciner-ation residues which are fine particles. Typically, bottom ash makes up approximately 20–25% by weight or 5 to 10% by volume of the original waste (AECOM report

2009). At most incineration facilities, bottom ash is mechanically collected, cooled (sometimes water quenched then drained), and mechanically, magnetically or electrically screened to recover recyclable metals. The remaining residue is typically disposed of at a landfill. The bottom ash may also be utilized as a substitute to construction aggregate, provided it has the right physical properties and chemical composition and meets regulatory requirements in the applicable field (EC 2006, AECOM report 2009). Its structural properties are appropriate to be used in road construction as a base or sub-base material (Forteza et al. 2004, Pfrang-Stotz and Reichelt 1997).

However, the bottom ash contains inorganic pollutants (which are not destroyed in the incineration process) such as metals and metalloids (e.g., Cu, Pb, Zn), including elements forming oxyanions (e.g., As, Cr, Mo, Sb) as well as easily soluble chloride salts and sulfates (Chandler et al. 1997). Therefore, the reuse of bottom ash as a construction material allows the risk of environmental pollution through leaching. In recent years, several studies were undertaken with the aim to understand and minimise the pollution potential of the incineration bottom ash (e.g., EC 2006, Vehlow 2002, Bruder-Hubscher et al. 2001, Cai et al. 2004, Comans et al. 2000, Forteza et al. 2004, Schreurs et al. 2000). But, no efficient energy and time saving treatment process for preventing environmentally harmful leaching from incineration residues utilised in construction and mine reclamation has yet recognized (Todorovic 2006).

The Stantec report (2011), however, typically classified bottom ash from MSW bass-burn facilities as a nonhazardous waste as the standard test methods have shown the constituents in the ash are not leachable. This probably indicates that contaminants are not mobile and are chemically/mechanically bound in the ash matrix. The disposal of bottom ash in a landfill or in using for beneficial purpose is therefore sound.

In Europe, MSW is increasingly pre-sorted before incineration. About 20 million metric tonnes per year of incinerator bottom ash is produced and used as a substitute for valuable primary aggregate resources in the construction of roads and embankments. Germany and France (the largest producers) along with Denmark and Netherlands use more than 60% of the bottom ash generated in road base, highway sound barriers, embankments, parking lots, bicycle paths and concrete and asphalt products. In Netherlands, virtually all incinerator residues are reused. In the UK, an increasing supply of high quality bottom ash has led to accepting it as a secondary aggregate with both environmental and cost benefits (Foth Infrastructure & Environment, 2013). European bottom ash contains approximately 2 million metric tonnes of iron scrap and 400,000 tonnes of non-ferrous metals, which corresponds to 1% of European iron and steel production and non-ferrous metals production (EU Statistics 2007). Recycling of bottom ash is very regular in Europe and metals recovery from the bottom ash is common for ferrous metals. The main non-ferrous metal is aluminium, which extends through the entire particle size range of the ash and the remaining metals are a mixture of copper, brass, zinc, lead, stainless steel and precious metals. Most of the precious metals exist as alloys from jewellery in the 2 to 6 mm fraction and as remainders of electronic components in the 0 to 2 mm fraction. The presence of precious metals in bottom ash in levels of 10 ppm of silver and 0.4 ppm of gold in European facilities has been acknowledged, apparently coming from incineration of electrical and electronic equipment for smaller particle sizes and from jewellery for the larger particle sizes. In Netherlands, aluminium, copper, silver and other metals worth $67.3 million per year have been recognized as lost through incineration.

Waste incineration bottom ash treatment involves 'deep dry recovery' of metal particles (>4 mm) by dry physical methods, followed by use of the residue in infrastructure as in Denmark; and 'wet treatment' of the bottom ash to remove organics and metals down to 0.1 mm, producing at the same time a clean aggregate for the building industry as in the Netherlands (Muchova et al. 2006, 2007, Bakkar et al. 2007, van Gerven 2005).

The bottom ash treatment concepts developed by Amsterdam and Indaver WtE Plants (5.1 and 5.10 in Chapter 5) aim at a more distinct separation without making any modification to the way bottom ash is being discharged from the combustion process. The processes are of a sort similar, consisting of sieving/washing steps, metal removal and flotation/density separation. As output both processes have ferrous and non-ferrous metals for recycling; granulates that are re-used in the cement or asphalt industry; a *sand* fraction that can either be used in the cement industry or used for landfill construction. And a sludge/filter cake which needs to be landfilled. More and cleaner metals are thus recovered and a substantial smaller portion cannot be re-used. Switzerland has focused on 'dry bottom ash treatment' where larger amounts of metals, especially non-ferrous, in an improved quality from the bottom ash is separated. The dry discharge of the bottom ash is essential for this, which avoids distributing the fine fraction, containing many heavy metals, over all materials, especially the metals. This has the additional benefit to save cost for water consumption, but mainly for transport and landfilling of about 15% water, typical for bottom ash from a wet discharging system (Fleck 2012).

Lund University in Sweden has reported an alternate technology to produce hydrogen using bottom ash, claimed to have the potential to produce up to 20 billion liters of hydrogen a year, or 56 GWh. The technique involves placing the ash in an oxygen-free environment, where when dampened with water it forms hydrogen gas. The gas is sucked up through pipes and stored in tanks. The results of hydrogen formation showed the commercial possibility of the hydrogen production from fresh bottom ash at mild temperatures and pressures. Compared to metal recovery, the reaction of metallic Al with alkaline solutions, to generate hydrogen was more efficient. The ash can be used as a resource through recovery of hydrogen instead of being released into the air as at present (Ilyas, Amir 2013). The research on the production hydrogen from MSWI bottom ash may also be found in Biganzoli et al. (2013) and Larsson (2014).

In the U.S., only 7% of the 7 million tons of bottom ash produced in the waste incinerators is utilized with the remaining 93% landfilled. A few States have approved for use as construction material either on a case-by-case or pre-approved basis. Several States have approved for various landfill uses such as daily cover. For example, Florida State has certified the bottom ash suitable for road construction, not subject to regulations for waste materials; and similarly a soil cement substitute made from ash. Florida and Massachusetts states had approved the use of ash in manufacture of asphalt (Foth Infrastructure & Environment 2013).

3.6.2 Treatment and use of APC residue

Since APC residues include the particulate matter captured after the acid gas treatment units, the residue may be a solid or a sludge, depending on the air pollution control equipment used (dry, semi-dry or wet processes).

Compared to bottom ash, APC residues are classified and managed as hazardous wastes as they typically contain elevated concentrations of heavy metals. Fly ash and APC residues are hazardous wastes because of mechanical and chemical behavior of the constituents in the emission. Fine particulate present in the flue gas has been found to form a nucleus on which volatilized metals evolved in the combustion zone condense (Chiang K.Y. et al. 1997). Like bottom ash, the composition of APC residues and of fly ash vary depending on the composition of the waste feed to the incinerator (Stantec 2011).

These residues generally have limited applications, mainly because they are characterized by large quantities of soluble salts such as sodium and potassium chlorides and calcium compounds, significant amounts of toxic heavy metals such as Pb, Zn, Cr, Cu, Ni, Cd, Hg in forms that may easily leach out, and trace amounts of toxic organics such as dioxins/furans. Therefore, the residues require pre-treatment to improve their environmental characteristics before landfill. The details of the constituents in the APC residues are given in Stantec report (2011). Several studies suggest two possible ways of handling these residues: (i) landfill after adequate treatment, or (ii) alternative uses (recycling as a secondary material). The most promising materials for recycling this residue are ceramics and glass–ceramic materials.

The treatment techniques may be grouped into separation processes, solidification/ stabilization, and thermal methods. These methods have advantages and disadvantages: lower costs are associated with the stabilization/solidification methods compared to the separation or thermal treatment options. Temporary storage in big bags and/or disposal in underground sites are alternatives undertaken in some countries.

Most of the studies on MSW ash have focused on the origin, behavior and mitigation of heavy metals, dioxins, furans and other toxic and carcinogenic materials found in the fly ash residue due to their toxic nature. As the ash has been preferred to put to beneficial uses rather than landfill, treatment options must also include other compounds beside heavy metals and carcinogens. For example, soluble chloride salts in the ash may not cause a threat to humans or the environment in landfill conditions; however they do pose a problem for use in building materials where they can catalyze material decay and/or in mine reclamation application where elevated chloride levels could be present in leachate that enters surface waters. Hence, if ash is considered for various reuse applications, a sound mitigation mechanism for chloride must be developed. The chloride salts are generally distributed between the bottom ash and the fly ash with a small fraction lost to stack emissions. Investigations have shown that roughly 60% of the total incoming chloride ends up in fly ash (Chang and Huang 2002). By focusing mitigation of chloride to fly ash, the efficiency of chloride treatment can be maximized while minimizing overall material handling. Adam Penque (2007) examines in detail the origins and the potential for treatment of chlorides in MSW combustion residue.

There are several publications on the composition, treatment and beneficial uses of fly ash/APC residues. For example, Quina et al. (2008) discussed several issues: the chemical constituents in these ashes using several techniques, treatment methods and environmentally sound usages under practice in different countries. Beneficial utilization of residues from MSW incineration is an important goal for integrated waste management (Van der Sloot et al. 2001).

The physical and chemical nature of the APC residues mainly depends on the type of combustion system and flue gas treatment methods (Alba et al. 1997).

Only a few studies regarding the physical and chemical nature of residues from fluidized bed combustion have been published. Blondin et al. (1997) and Abbas et al. (1999, 2001, 2003) have investigated the chemical composition and leaching behavior of both bottom ash and APC residues from BFB boilers. In practice, the most common options for the management of fly ash/APC residues are permanent storage in hazardous waste disposal sites or treatment followed by disposal.

Nine potential applications were identified for incineration residues (Ferreira et al. 2003) which can be grouped into four main categories: construction materials (cement, concrete, ceramics, glass and glass–ceramics); geotechnical applications (road pavement, embankments); agriculture (soil amendment), and others (sorbent, sludge conditioning).

In addition, several components present in fly ash may be recovered and reused. The primary interest is centered on the recovery of salts, acid, gypsum, and metals: (a) Salt recovery directly from the residues is possible after water extraction. This has been considered in conjunction with several treatment technologies generating salt containing process water. This technique is in commercial use today; (b) acid: the solution from a first scrubber stage of a multi-stage APC setup is essentially concentrated hydrochloric acid. Techniques to recover this acid are in commercial use; (c) production of gypsum can be achieved based on recovery of gypsum from the scrubber solution from alkaline scrubbers. This technique is in commercial use; and (d) metals can be recovered using extraction and thermal techniques. This technique is in commercial use (ISWA 2008).

Romero et al. (2001) identify several applications for this waste, particularly building materials, catalysts, refractories, pozzolanic materials, glasses and glass-ceramics. Mohapatra and Rao (2001) reviewed aspects related to the use and environmental effects of fly ashes in general, resulting from the burning of municipal solid waste and other fuels.

Application of APC residue in ceramic and glass-ceramics (Talmy et al. 1996), synthetic aggregates (Wainwright and Cresswell 2001), and in eco-cement production (Ampadu and Torii 2001) were also investigated.

Reliable results in terms of treatment and application may be obtained by combining two or more methods, e.g., washing for removing salts followed by solidification/ stabilization or thermal methods. Although washing can generate additional environmental problems, it results in a material that is suitable for further treatment or may have practical applications.

The production of lightweight aggregates by incorporating APC residues with natural clays shows promise.

In order to avoid landfill, the valorization of the residues has been studied and several applications have been tested. Applications in constructing materials, geotechnical applications, agriculture and others are identified and investigated.

In summary, APC residues are in general considered hazardous waste, which forbids their usage directly as raw material in environmental applications, and should be treated before landfill.

REFERENCES

Abbas, Z., Andersson, B.A. & Steenari, B.M. (1999) Leaching behavior and possible resource recovery from air pollution control residues of fluidized bed combustion of municipal solid waste. In: *Proc. of 15th International Conference on Fluidized Bed Combustion, 16–19 May.* Savannah, USA, ASME.

Abbas, Z., Lindqvist, O. & Steenari, B.M. (2001) A study of Cr(VI) in ashes from fluidized bed combustion of municipal solid waste: Leaching, secondary reactions and the applicability of some speciation methods. *Waste Management*, 21, 725–739.

Abbas, Z., Moghaddam, A.P. & Steenari, B.-M. (2003) Release of salts from municipal waste combustion residues. *Waste Management*, 23, 291–305.

AEA Technology plc (January 3, 2012) *Review of Research into Health Effects of Energy from Waste Facilities.* Report for Environmental Services Association; Author: Broomfield, M.

AECOM Canada Ltd. (2009) *Management of Municipal Solid Waste in Metro Vancouver—A Comparative Analysis of Options for Management of Waste After Recycling.*

A.J. Chandler & Associates Ltd (2006) *Review of Dioxins and Furans from Incineration.* A Report prepared for 'The Dioxins and Furans Incineration Review Group', Canadian Council of Ministries of the Environment Inc., 2007.

Alba, N., Gasso, S., Lacorte, T. & Baldasano, J.M. (1997) Characterization of municipal solid waste incineration residues from facilities with different air pollution control systems. *Journal of Air and Waste Management Association*, 47, 1170–1179.

Ampadu, K.O. & Torii, K. (2001) Characterization of ecocement pastes and mortars produced from incinerated ashes. *Cement Concrete Research*, 31, 431–436.

Aurell, J. (2008) *Effects of Varying Combustion Conditions on PCDD/F Formation.* Doctoral Thesis. Umeå, Sweden, Chemistry Department, Umeå University.

Bakker, E.J., Muchová, L. & Rem, P.C. (2007) Separation of precious metals from MSWI bottom ash. In: *Conf. Proceedings, 6th International Industrial mineral Symposium, 1–3 February, Izmir, Turkey.* p. 6.

Beckmann, M. & Scholz, R. (2000) Residence time behavior of solid material in Grate systems. In: *5th Europe Conf. Ind. Furnaces and Boilers INFUB, Porto, Portugal, 11./14./4.2000.*

Bergström, J.G.T. & Warman, K. (1987) Production and characterization of trace organic emissions in Sweden. *Waste Management & Research*, 5, 395–401.

Biganzoli, L., Ilayas, A., van Praagh, M., Persson, K.M. & Grosso, M. (2013) Aluminium recovery vs hydrogen production as resource recovery options for fine MSWI bottom ash fraction. *Waste Management*, 33, 1174–1181.

Bilitewski, B., Härdtle, G. & Marek, K. (1997) *Waste Management.* Berlin, Springer. ISBN: 3-540-59210-5.

Blondin, J., Iribarne, A.P. & Anthony, E.J. (1997) Study of ashes from municipal wastes incinerated in classical grate boilers and FBC units: Solidification of MWIR with FBC coal ashes for disposal. In: *Proc. of 14th International Conf. on Fluidized Bed Combustion, ASME.*

Blumenstock, M., Zimmermann, R., Schramm, K.-W. & Kettrup, A. (2000) Influence of combustion conditions on the PCDD/F-, PCB-, and PAH-concentrations in the post-combustion chamber of a waste incineration pilot plant. *Chemosphere*, 40, 987–993.

BREF (2006) *Integrated Pollution Prevention and Control—Reference Document on the Best Available Techniques for Waste Incineration.* European Commission.

BREF Report (December 2001) Draft of a German Report with basic information for a BREF-document "Waste Incineration". BREF report english—FTP Direct listing. Available from: files.gamta.It/aaa/Tipk/tipk/4_kiti%20GPGB/63.pdf.

Bruder-Hubscher, C., Lagarde, F., Leroy, M.J.F., Couganowr, C. & Enguehard, F. (2001) Utilisation of bottom ash in road construction: Evaluation of the environmental impact. *Waste Management Research*, 19 (6), 545–556.

Bunsan, S., Chen, W.Y., Chen, H.W., Chuang, Y.H. & Grisdanurak, N. (2013) Modeling the dioxin emission of a municipal solid waste incinerator using neural networks. *Chemosphere*, 92 (3), 258–264.

Cai, Z., Bager, D.H. & Christensen, T.H. (2004) Leaching from solid waste incineration ashes used in cement-treated base layers for pavements. *Waste Management*, 24, 603–612.

Chandler, A.J., Eighmy, T.T., Hartlen, J., Hjelmar, O., Kosson, D.S., Sawell, S.E., Sloot, H.A.v.d. & Vehlow, J. (1997) *Municipal Solid Waste Incinerator Residues*. Amsterdam, Elsevier Science B.V.

Chiang, K.Y., Wang, K.S. & Lin, F.L. (1997) The effect of inorganic chloride on the partitioning and speciation of heavy metals during a simulated municipal solid waste incineration process. *Toxicological & Environmental Chemistry*, 64 (1–4).

Comans, R.N.J., Meima, J.A. & Geelhoed, P.A. (2000) Reduction of contaminant leaching from MSWI bottom ash by addition of sorbing components. *Waste Management*, 20 (2–3), 125–133.

Environment Australia (1999) *Incineration and Dioxins: Review of Formation Processes*. Consultancy report prepared by Environmental and Safety Services for Environment Australia, Commonwealth Department of the Environment & Heritage, Canberra.

European Commission (August 2006) *Integrated Pollution Prevention and Control, Reference Document on the Best Available Techniques for Waste Incineration*.

EU Statistics, (2007) *Structural Indicators: Municipal Waste*. Available from: http://Europe. eu.int.

Fängmark, I., van Bavel, B., Marklund, S., Strömberg, B., Berge, N. & Rappe, C. (1993) Influence of combustion parameters on the formation of polychlorinated dibenzo-*p*-dioxins, dibenzofurans, benzenes, and biphenyls and polyaromatic hydrocarbons in a pilot incinerator. *Environmental Science & Technology*, 27, 1602–1610.

Feasibility Study of Thermal Waste Treatment/Recovery Options in the Limerick/Clare/Kerry Region Ireland, August 2005; MDE0267Rp0003 32 Rev F01; Chapter 4. Thermal Waste Treatment Technologies. p. 38.

Ferreira, C., Ribeiro, A. & Ottosen, L. (2003) Possible applications for municipal solid waste fly ash. *Journal of Hazardous Materials*, 96 (2–3), 201–216.

Fleck, E. (October 2012) *Waste Incineration in 21st Century—Energy Efficient and Climate-Friendly Recycling Plant & Pollutant Sink*.

Forteza, R., Far, M., Segui, C. & Cerda, V. (2004) Characterization of bottom ash in municipal solid waste incinerators for its use in road base. *Waste Management*, 24 (9), 899–909.

Foth Infrastructure & Environment, LLC (July 2013) *Alternative Technologies for Municipal Solid Waste*. Report prepared for Ramsey Washington County Resource Recovery Project.

GEA Brochure. *Spray Drying Absorption—Easy Way to Clean the Flue Gas from Waste Incinerators*. Denmark, GEA Niro. Available from: www.niro.com/niro/cmsdoc.nsf/WebDoc/ ndkw6ptc4z.

GENIVAR Ontario Inc. in association with Ramboll Danmark A/S (2007) *Municipal Solid Waste Thermal Treatment in Canada*.

German Federal Environment Agency (GFEA) (2001) Draft of a German Report for the creation of a BREF-document "Waste Incineration".

Gullett, B.K. & Raghunathan, K. (1997) Observations on the effect of process parameters on dioxin/furan yield in municipal waste and coal systems. *Chemosphere*, 34, 1027–1032.

Gullett, B.K., Dunn, J.E., Bae, S.-K. & Raghunathan, K. (1998) Effects of combustion parameters on polychlorinated dibenzodioxin and dibenzofuran homologue profiles from municipal waste and coal co-combustion. *Waste Management*, 18, 473–483.

Hartenstein, H.-U. & Licata, A. (2008) *Modern Technologies to Reduce Emissions of Dioxins and Furans from Waste Incineration, Nawtec08*. Available from: http://www.seas. columbia.edu/earth/wtert/sofos/nawtec/nawtec08/nawtec08-0010.pdf.

Hites, R.A. (2011) Dioxins: An overview and history. *Environmental Science & Technology*, 45, 16–20.

Huang, H. & Buekens, A. (2000) Chlorinated dioxins and furans as trace products of combustion: Some theoretical aspects. *Toxicological & Environmental Chemistry*, 74, 179–193.

IEA Bioenergy. *Accomplishments from IEA Bioenergy-Task 36: Integrating Energy Recovery into Solid Waste Management Systems (2007–2009), End of Task Report.* Chapter 4.

Ilyas, A. (February 18, 2013) *Unsaturated Phase Environmental Processes in MSWI Bottom Ash.* Dissertation. Sweden, Department of Water resources Engineering, Lund University.

ISWA (2008) *Management of APC Residues from WTE Plants.*

Jansson, S. (2008) *Thermal Formation and Chlorination of Dioxins and Dioxin-Like Compounds.* Doctoral Dissertation. Sweden, Department of Chemistry, Umea University. Available from: www.diva-portal.org/smash/get /diva2:142298/FULLTEXT01.pdf [Retrieved 21th February 2014].

Jungten, H., Richter, E., Knoblauch, K. & Hoang-Phou, T. (1988) Catalytic NO_x reduction by ammonia and carbon catalysts. *Chemical Engineering Science*, 43, 419–428.

Kilgroe, J.D., Nelson, L.P., Schindler, P.J. & Lanier, W.S. (1990) Combustion control of organic emissions from municipal waste combustors. *Combustion Science and Technology*, 74, 223–244.

Kilgroe, J.D., Lanier, W.S. & Von Alten, T.R. (1992) Development of good combustion practice for municipal waste combustors. In: *Proceedings of the National Waste Processing Conference.* Vol. 15. pp. 145–156.

Klasen, T. & Gorner, K. (1999) Numerical calculation and optimization of a large municipal waste incinerator plant. In: *2nd Intl. Symposium on Incineration and Flue Gas Treatment Technologies.* Sheffield, UK, Sheffield University.

Kokalj, F. & Samec, N. Combustion of municipal solid waste for power production. *INTECH (Open Science/Open Minds).* doi:10.5772/55497.

Korell, J., Andersson, S., Löthgren, C.J., Seifert, S. & Paur, H.-R. (2006) *Simultane Abscheidung von Quecksilber, Dioxinen und Feinstpartikeln aus Rauchgasen.* Paper presented at VDI-GVC-Fachauschuss Energieverfahrenstechnik, Würzburg/Germany.

Lam, C.H.K., Ip, A.W.M., Barford, J.P. & Mckay, G. (2012) Use of MSW incineration ash: A review. *Sustainability*, 2, 1943–1968, Open Access.

Larsson, R. (2014) *Energy Recovery of Metallic Al in MSWI Bottom Ash, Different Approaches to Hydrogen Production from MSWI Bottom Ash: A Case Study.* MS Thesis in Energy Engineering. Umea, Sweden, Tekniska Hogskolan, Umea University. Available from: www.diva-portal.org/smash/get/diva2:757224/FULLTEXT01.pdf.

Lopes, E.J., Okamura, L.A. & Yamamoto, C.I. (2015) Formation of dioxins and furans during MSW gasification. *Brazilian Journal of Chemical Engineering*, 32 (1), Sao Paulo, January/March 2015.

Mätzing, H., Baumann, W., Becker, B., Jay, K., Paur, H.-R. & Seifert, H. (2001) Adsorption of PCDD/F on MWI fly ash. *Chemosphere*, 42, 803–809.

McDougall, F., Thomas, B. & Dryer, A. (2002) Life cycle assessment for sustainable solid waste management—An introduction. *Waste Management*, May 2002, 43–45.

McDougall, F., While, P., Franke, M. & Hindle, F. (2002) *Integrated Solid Waste Management: A Life Cycle Inventory.* Oxford, UK, Blackwell.

McKay, G. (2002) Dioxin characterization, formation and minimization during municipal solid waste incineration: A review. *Chemical Engineering Journal*, 86, 343–368.

Mohapatra, R. & Rao, J.R. (2001) Review—Some aspects of characterisation, utilisation and environmental effects of fly ash. *Journal of Chemical Technology and Biotechnology*, 76, 9–26.

Muchová, L. & Rem, P. (2006) Pilot plant for wet physical separation of MSWI bottom ash. In: *Conf. Proc. MMME 2006, Cape Town (South Africa), 14–15 November.* p. 12.

Muchová, L., Rem, P. & Van Berlo, M. (2007) Innovative technology for the treatment of bottom ash. In: *Conference Proceeding from ISWA/NVRD World Congress 2007, Amsterdam, The Netherlands, 24–27 September 2007.*

Olie, K., Vermeulen, P. & Hutzinger, O. (1977) Chlorodibenzo-p-dioxins and chlorodibenzo furans are trace components of fly ash and flue gas of some municipal incinerators in The Netherlands. *Chemosphere*, 6, 455–459.

Penque, A. (September 2007) *Examination of Chlorides in Municipal Solid Waste-to-Energy Combustion Residue: Origins, Fate and Potential for Treatment*. Thesis submitted for the Master's degree in Earth Resources Engineering. New York, NY, Department of Earth and Environmental Engineering, Columbia University.

Pfrang-Stotz, G. & Reichelt, J. (1997) Municipal solid waste incineration (MSWI) bottom ashes as granular base material in road construction. In: *Waste Materials in Construction—Putting Theory into Practice*. Amsterdam, The Netherlands, Elsevier Science. pp. 85–90.

Quina, M.J., Bordado, J.C.M. & Quinta-Ferreira, R.M. (2008) Treatment and use of air pollution control residues from Municipal Solid Waste Incineration: An overview. *Waste Management*, 28, 2097–2121.

Quina, M.J., Bordado, J.C.M. & Quinta-Ferreira, R.M. (2011) Air pollution control in municipal solid waste incinerators. In: Khallaf, M. (ed.) *The Impact of Air Pollution on Health, Economy, Environment and Agricultural Sources*. InTech. ISBN: 978-953-307-528-0. Available from: http://www.intechopen.com/books/the-impact-of-air-pollution-on-health-economy-environment-andagricultural-sources/air-pollution-control-in-municipal-solid-waste-incinerators.

Romero, M., Rincón, J.M., Rawlings, R.D. & Boccaccini, A.R. (2001) Use of vitrified urban incinerator waste as raw material for production of sintered glass–ceramics. *Materials Research Bulletin*, 36, 383–395.

Sabbas, T., Polettini, A., Pomi, R., Astrup, T., Hjelmar, O., Mostbauer, P., Cappai, G., Magel, G., Salhofer, S., Speiser, C., Heuss-Assbichler, S., Klein, R. & Lechner, P. (2003) Management of MSW incineration residues. *Waste Management*, 23, 61–88.

Schreurs, J.P.G.M., Sloot, H.A.v.d. & Hendriks, C. (2000) Verification of laboratory—Field leaching behavior of coal fly ash and MSWI bottom ash as a road base material. *Waste Management*, 20 (2–3), 193–201.

Section 4: Emission Estimates. Available from: www.dep.state.fl.us/air/emission/construction/hillsborough/VolumeIIISEC4.pdf.

Stantec (March 2011) *Waste to Energy: A Technical Review of Municipal Solid Waste Thermal Treatment Practices*—Final Report.

Talmy, I.G., Haught, D.A. & Martin, C.A. (1996) *Ash-Based Ceramic Materials*. US Patent no. 5521132.

Thome-Kozmiensky, K.J. (1994) *ThermischeAbfallbehandlung*. EF-Verlag fuer Energie-und Umwelttechnik GmbH.

Todorovic, J. (2006) *Pre-treatment of Municipal Solid Waste Incineration (MSWI) Bottom Ash for Utilization in Road Construction*. Doctoral Thesis. Lulea, Sweden, Division of Waste Science and Technology, Department of Civil and Environmental Engineering, Lulea University of Technology.

Todorovic, J. & Ecke, H. (2006a) Demobilisation of critical contaminants in four typical waste-to-energy ashes by carbonation. *Waste Management*, 26 (4), 430–441.

Todorovic, J. & Ecke, H. (2006b) Treatment of MSWI residues for utilization as secondary construction minerals: A review of methods. *Minerals and Energy—Raw Materials Report*, 20 (3), 45–59.

TWGComments (2003) *TWG Comments on Draft 1 of Waste Incineration BREF*.

UNEP Chemicals (2005) *Standardized Toolkit for Identification and Quantification of Dioxin and Furan Releases*. Genebra, Switzerland.

U.S. EPA (10/96) *Solid Waste Disposal, 2.1 Refuse Combustion—EPA*. Available from: www.epa.gov/ttnchie1/ap42/ch02/final/c02s01.pdf.

U.S. EPA (October 2003) *Municipal Solid Waste in the United States: 2001 Facts and Figures.* Office of Solid Waste and Emergency Response. EPA530-R-03-011.

U.S. EPA (2005a) *Treat, Store, and Dispose of Waste.* Available from: http://www.epa.gov/epaoswer/osw/tsd.htm.

U.S. EPA (2005b) *Endocrine Disrupting Chemicals Risk Management Research.* Available from: http://www.epa.gov/ORD/NRMRL/EDC/.

U.S. EPA (2006) *Combustion Emissions from Hazardous Waste Incinerators, Boilers and Industrial Furnaces, and Municipal Solid Waste Incinerators—Results from Five STAR Grants and Research Needs.* Washington, DC, Office of Research and Development, National Center for Environmental Research, EPA/600/Q-06/002.

Van der Sloot, H.A., Kosson, D.S. & Hjelmar, O. (2001) Characteristics, treatment and utilization of residues from municipal waste incineration. *Waste Management,* 21 (8), 753–765.

Van Gerven, T. (2005) Management of incinerator residues in Flanders (Belgium) and in neighbouring countries: A comparison. *Waste Management,* 25, 75–87.

Vehlow, J. (2002) Bottom ash and APC residues management. In: *Power production from Waste and Biomass IV, Helsinki.* 951-38-5734-4.

Vehlow, J. (2012) *Overview of Waste Incineration Technologies.* IEA Bioenergy Workshop ExCo 71, May 21, 2012, Cape Town, South Africa.

Vehlow, J. (2015) Air pollution control systems in WtE units: An overview. *Waste Management,* 37, 58–74.

Violi, A., D'Anna, A. & D'Alessio, A. (2001) A modeling evaluation of the effect of chlorine on the formation of particulate matter in combustion. *Chemosphere,* 42, 463–471.

Vogg, H., Metzger, M. & Stieglitz, L. (1987) Recent findings on the formation and decomposition of PCDD/PCDF in MSW incineration. *Waste Management & Research,* 5, 285–294.

Wainwright, P.J. & Cresswell, D.J. (2001) Synthetic aggregates from combustion ashes using an innovative rotary kiln. *Waste Management,* 21, 241–246.

Waste-C-Control. *Data Base of Waste Management Technologies—Grate Incineration Technology.* Available from: www.epem.gr/waste-c control/database/html/WtE-01.htm [Retrieved 2th April 14].

Werther, J. (2007) Gaseous emissions from waste combustion. *Journal of Hazardous Materials,* 144, 604–613. Available from: www.aseanenvironment.info/abstract/41015669.pdf.

Werther, J. & Ogada, T. (1999) Sewage sludge combustion. *Progress in Energy and Combustion Science,* 25 (1), 55–116.

Wikström, E., Marklund, S. & Tysklind, M. (1999) Influence of variation in combustion conditions on the primary formation of chlorinated organic micropollutants during municipal waste combustion. *Environmental Science & Technology,* 33, 4263–4269.

World Bank (1999) *Decision Maker's Guide to Municipal Solid Waste Incineration.* Washington, DC, The International Bank for Reconstruction and Development/The World Bank.

WSP (2013) *Review of State-of-the-Art Waste-to-Energy Technologies, Stage 2—Case Studies.* Prepared by Kevin Whiting, WSP Environmental Ltd, WSP House, London, UK.

WtERTC (August 2014) *Waste Incineration—Flue Gas Cleaning.* Waste-to-Energy Research and Technology Council, created by Prof. Dr.-Ing. Rudi Karpf. Available from: http://www.wtert.eu/default.asp?Menue=12&ShowDok=13.

Yi, J., Sauer, H., Leuschke, F. & Baege, R. (2005) What is possible to achieve on flue gas cleaning using the CFB technology? In: Cen, K. (ed.) *Circulating Fluidized Bed Technology VIII.* Beijing, International Academic Publishers. pp. 836–843.

Zakaria, R., Goh, Y., Yang, Y., Lim, C., Goodfellow, J., Chan, K., Reynolds, G., Ward, D., Siddall, R., Naasserzadeh, V. & Swithenbank, J. (2000) Fundamental aspects of emissions from the burning bed in a MSW Incinerator. In: *5th European Conf. Ind. Furnaces and Boilers INFUB, Porto, Portugal, 11./14./4.2000.*

Zimmermann, R., Blumenstock, M., Heger, H.J., Schramm, K.-W. & Kettrup, A. (2001) Emission of nonchlorinated and chlorinated aromatics in the flue gas of incineration plants during and after transient disturbances of combustion conditions: Delayed emission effects. *Environmental Science & Technology*, 35, 1019–1030.

Chapter 4

Pyrolysis and Gasification technologies

4.1 INTRODUCTION

Pyrolysis and Gasification are thermal processes developed as alternatives to combustion process for treating municipal wastes to recover energy from its organic fraction. These are advanced thermal technologies involving more complex processes, and offer better prospect in utilizing the waste than landfilling and combustion. However, their commercial experience so far is not adequate (Chen et al. 2014, Bosmans et al. 2012, Di Gregorio et al. 2012, RTI International 2012, Castaldi 2008, Castaldi and Themelis 2010, E4Tech 2009, Arena 2012, Ricketts et al. 2002, Morris and Waldheim 1998, Manya et al. 2006, Kim 2003, Zhao et al. 2010, Yassin et al. 2009, Arena et al. 2010 & 2011, Xiao et al. 2009, Thomas 2004, ESTET 2004, Belgiorno et al. 2003, Surisetty et al. 2012, Faaij et al. 1997, Niessen et al. 1996, Morris and Waldheim 1998, Solantausta et al. 1999).

Pyrolysis and gasification have the ability to produce a consistent, high-quality syngas that can be used for producing energy in the form of heat and electricity. It can also be used as a feedstock to produce high value chemicals and fuels.

Pyrolysis and gasification for energy recovery from waste are, in most cases, followed by immediate combustion of their gases, and in the case of pyrolysis, also solid reaction products.

The nature and design of these systems prevents oxidation of the feedstock leading to the formation of fewer pollutants that can be removed significantly. As a result, these technologies comply even with the most stringent regulatory requirements of any country.

4.2 PYROLYSIS

Pyrolysis is an endothermic thermochemical conversion of carbonaceous material in the absence of air/oxygen or with such a restricted supply that gasification occurs to a limited extent. This partial gasification is used to provide the thermal energy required for pyrolysis; but, at the expense of the product yields which are syngas, liquids (pyrolytic oil), and solid char. The process occus at low temperatures, 400–900°C, but usually lower than 700°C (Helsen and Bosmans 2010). The yield and composition of the products is very much dependent on the type of materials, pyrolysis method and the process conditions such as the temperature, pressure, reaction rates,

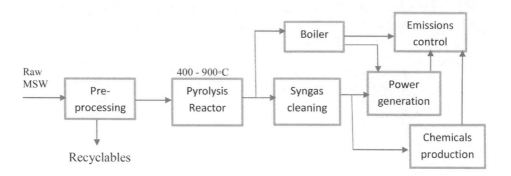

Figure 4.1 Flow diagram of pyrolysis of solid waste.

heat transfer rate, the gas residence time, the presence of catalysts, and the presence of hydrogen donors, such as hydrogen and steam vapor. The pyrolysis flow diagram is shown in Figure 4.1.

Some of the waste materials are composed of complex chemical polymers of long chains, and pyrolyzing such materials in the absence of oxygen breaks up the long chains to produce chains and molecules with lower molecular weights. These shorter molecules result in the formation of the oils and hydrocarbon gases including hydrogen gas, characteristic of pyrolysis of waste.

The main constituents of pyrolysis gas are CO, H_2 and CH_4, which are combustible gases. Also contains oxidized compounds (carbon dioxide and water) which have no heating value, but dilute the gas. As the pyrolysis process occurs at low temperatures with no air (oxygen), a large unreacted portion of the feedstock remain in the form of solid char.

Pyrolysis can be a self-standing treatment, but is mostly followed by combustion or in some cases, liquifqction. The heating values of pyrolysis gas is typically in the range, 5–15 MJ/m^3 based on MSW and 15–30 MJ/m^3 based on RDF (BREF report, 2001).

Due to the lower temperatures and reducing conditions, the pyrolysis of MSW has the advantage of reducing and preventing corrosion and emissions by retaining alkali, heavy metals, chlorine and sulfur within the process residues, preventing dioxin and furan formation, and reducing thermal NO_x formation (Liu and Liu 2005).

The pre-processing of MSW is typically required for the process to be more efficient. Depending on the specific pyrolysis process, pre-processing may include sorting and separation of themally non-degradable materials like glass, metals and concrete, followed by size reduction and densification.

In case, the waste has high moisture content, a dryer may be added to the pre-processing stage to lower the moisture to <25% so that the heating value of feedstock can be increased enabling the system to become more efficient. The waste heat or fuel produced by the system can be used to dry the incoming MSW.

The pyrolytic reactions can be slow or fast or conventional depending on the process conditions: slow pyrolysis with long residence times (hours to days) at 400°C produces mainly charcoal; conventional pyrolysis involving moderate heating rates

Table 4.1 Types of Pyrolysis technologies (Williams 2012).

Technology	Residence time	Heating rate	Temp. °C	Major products
Slow Pyrolysis (carbonization)	hours to days	Very low	400	Charcoal
Conventional Pyrolysis	5–30 min	Medium	400–600	Char, liquids & syngas
	5–30 min	Medium	700–900	Char & syngas
Fast Pyrolysis	0.1–2 sec	High	400–650	Liquids
	<1 sec	High	650–900	Liquids & syngas
	<1 sec	Very high	1000–3000	Syngas

(20°C/min) and temperatures typically in the range, 400–600°C produces a fairly even distribution of char, oil and gas depending on the nature of raw waste; in fast or flash pyrolysis, waste decomposes very quickly to generate mostly vapors and aerosols and charcoal and gas (Bridgwater 2012); and short residence times (<1 sec) and high heating rates with rapid cooling of pyrolysis gases results in the production of liquid at 400 to 650°C, liquid and syngas at 650 to 900°C, and syngas at 1000 to 3000°C as shown in Table 4.1 (Williams 2012).

Thus, process conditions can be optimized to produce a desired product. For example, liquid products (pyrolytic oils) are produced by lower pyrolysis temperatures while syngas is produced by higher pyrolysis temperatures.

It must be noted, the terms – slow, fast and flash pyrolysis – are subjective and have no precise definition of the residence times and heating rates; sometimes both fast pyrolysis and flash pyrolysis are characterized as fast pyrolysis.

Process

Waste is tipped into a bunker where a crane mixes the incoming material and moves it to a shredder and from here to another bunker. The mixed waste is then fed into a gas tight hopper arrangement by a screw- or piston feeder. The coarsely shredded waste then enters a reactor, normally an externally heated rotary drum operated under atmospheric pressure (inert or nitrogen atmosphere) with absence of oxygen. Inside the reactor, the waste is dried at 100–200°C, followed by (a) initiation of the decomposition of H_2S, H_2O and CO_2 at 250°C, (b) breaking of the bonds of aliphatic substances (start of the separation of CH_4 and other aliphatics) at 340°C, (c) enrichment of the produced material in carbon (formation of pyrolysis coke) at 380°C, (d) breaking of C–O and C–N bonds at 400°C, (e) conversion of coal tar materials into fuel (pyrolysis oil) and tar at 400–600°C, (f) decomposition to materials resistant to heat, formation of aromatics at 600°C, and (g) production of aromatics, and processes for H_2 removal from organics like butadiene, cyclohexane etc. at >600°C (Moustakas and Loizidou 2010).

The thermal energy is usually applied indirectly by thermal conduction through the walls of a containment reactor since air or oxygen is not intentionally introduced or used in the reaction. The transfer of heat from the reactor walls occurs by filling the reactor with inert gas which also provides a transport medium for the removal of gaseous products.

Table 4.2 Pyrolysis reactors.

Reactor type	Heating method	Heating rate
Fluidized bed	Heated recycle gas	High
	Firetubes	Moderate
Entrained flow	Recycled hot sand	High
Fixed bed	Heated recycle gas	Low
Rotary kiln	Wall heating	Low

The preparation and grinding of the feedstock is very vital to improve and standardize the quality of the waste and to promote heat transfer. Also, depending on the process, a separated drying improves the lower heating value of the raw process gases and increases efficiency of gas-solid reactions within the pyrolysis reactor.

4.2.1 Pyrolysis reactors

Typically, pyrolysis systems use a drum, kiln-shaped tube, which is heated externally using either recycled syngas or a fuel such as natural gas. Since pyrolysis occurs in the absence of oxygen, the feed system and pyrolysis chamber are sealed and isolated from outside air during the processing. This is accomplished through the use of inlet and outlet knife-gates.

Other types of reactors are also used: entrained-flow, fixed-bed, cyclone gasifier, fluidized bed, plasma furnace and so on (e.g, IEA Bioenergy Task 36, Luo et al. 2010, Li et al. 2002, Garcia et al. 1995, Bosmans 2010), shown in Table 4.2.

The interaction between large numbers of thermochemical phenomena results in a huge diversity of substances and increases the complexity of the process. Several hundred different compounds are produced during waste pyrolysis, and many of these have not yet been identified. A thorough understanding of the characteristics and concentration of effluents to be processed is essential, especially when hazardous substances are concerned (Helsen 2000).

Pyrolysis gas has a complex composition and direct use requires extensive gas cleaning. This is especially difficult for the removal of sulfur compounds and dust particles or tar. That is why pyrolysis application in waste treatment is performed with direct combustion of the gas phase without major prior cleaning. The advantage of pyrolysis is the presence of good quality metal scrap in the solid residues which have clean metallic surfaces due to the absence of oxygen (IEA Bio-energy Task 36).

There are two other types, in addition to thermal pyrolysis: Catalytic pyrolysis/cracking and Hydrocracking (or hydrogenation).

In the catalyticpyrolysis, the feedstock is processed using a catalyst. The presence of a catalyst reduces the reaction temperature and time compared to thermal pyrolysis. The catalysts usually used are acidic materials, zeolites or alkaline compounds. Several investigations have shown that this method can be particularly used to process a variety of plastics including low density polyethylene, high density polyethylene, polypropylene and polystyrene (e.g. Ates et al. 2013, Miskolczi et al. 2013, Adrado et al. 2012, He et al. 2010).

In the hydrocracking, the feedstock reacts with hydrogen and a catalyst under moderate temperatures and pressures (around 150–400°C, and 30–100 bar hydrogen). This process was found highly appropriate to generate gasoline fuels from feedstocks like waste plastics, plastics mixed with coal, plastics mixed with refinery oils and scrap tyres (RTI Intl. 2012).

4.2.2   Investigations on pyrolysis of MSW

Pyrolysis of MSW was investigated by several research groups, and a few of them are outlined. Li et al. (1999) studied experimentally the pyrolysis of municipal solid waste in a laboratory-scale rotary kiln. The effects of heating methods, moisture content and size of waste on pyrolysis gas yields and compositions were evaluated. A rapid heating rate comparatively enhances gas production than a slow heating rate. Wood chips and polyethylene plastics present similar trends of heating values for the pyrolysis gases. The pyrolysis gases had high HHV levels. Mohammad et al. (2005) investigated the pyrolysis of municipal solid waste in a fixed-bed reactor to analyze pyrolytic oil. The results show that the liquid yield increased with an increase in temperature below 450°C. The collected liquid is highly oxygenated when the oxygen content is 52.91%, compared with 53.15% for raw feedstock. The carbon and hydrogen concentrations are similar to the carbon and hydrogen concentrations of biomass-derived oil. Buah et al. (2007) observed that the oil resulted from pyrolysis of municipal solid waste contained carboxylic acids and their derivatives: alkanes, alkenes, and mono and polycyclic and substituted aromatic groups. Further, high temperatures enhanced the yield of aromatic groups and lowered the yield of aliphatic groups. Velghe et al. (2011) studied the pyrolysis process for the production of valuable products from MSW in a semi-batch reactor. The presence of long aliphatic HCs revealed incomplete breakdown in the conditions of high heating rates and short residence times. A liquid with the lowest water content, lowest oxygen-to carbon ratio, and maximum yield was obtained by fast pyrolysis at 510°C. The oil fraction was abundant with 63.5% of aliphatic compounds and 23.5% of aromatic compounds. No waxy material was obtained in the liquid fractions produced by slow pyrolysis up to 550°C. The liquid product consisted of a water phase and an oil phase. The oil yield was low with low water content, low O/C value, and 70.2% aliphatic compounds.

More results can be obtained from the publications of, for example, Islam et al. (2005), Jiao et al. (2009), Saffarzadeh et al. (2009 a & b), Luo et al. (2010), Helsen and Bosmans (2010), He et al. (2010), Miskolczi et al. (2011), Li et al. (2011), Fonts et al. (2009 & 2012), Bosmans and Helsen (2012), Zhou (2013 & 2014). These studies show that pyrolysis technology has a high potential to thermally treat MSW to generate power as well as to derive value-based chemicals and liquid fuels. Malkow (2004) has discussed several novel and innovative pyrolysis technologies as applied to MSW. Chen et al. (2014) have recently published an excellent review of the pyrolysis technologies available for MSW that includes waste plastics.

Applicability: Besides thermally treating MSW and sewage sludge, pyrolysis is used for treating synthetic wastes and used tires, decontamination of soils, plastic waste and so on.

4.2.3   Plusses and minusses of the process

Plusses: The decomposition temperature is lower than the incineration temperature, so the thermal distress of the facility is less intense than in incineration; the decomposition takes place in reducing atmosphere unlike in incineration, and hence the demand for less oxygen results in less air emissions; the ash content in carbon is much higher than in the case of incineration; the metals that are included in waste are not oxidized during pyrolysis and have higher commercial value; the retention of heavy metals in the char is better than in ash from incineration process (at process temperature of 600°C, the retention of heavy metals: 100% chromium, 95% copper, 92% lead, 89% zinc, 87% nickel and 70% cadmium); no ash is produced from the combustion of the pyrolysis gas and the cleaning of the resulting gas is simpler; the initial waste volume is highly reduced in comparison with the incineration; product gas with LCV of 8 MJ/kg may be combusted in a compact combustion chamber with short retention time, and the emissions after extensive flue gas cleaning are low; CO_2 neutral energy production; HCl can be retained in or distilled from the solid residue; no formation of dioxins and furans; well suited for waste fractions with a high volatile content such as plastic wastes; high power to heat ratio, if co-fired in existing boiler plants with an efficient steam cycle, but low overall energy efficiency.

Minuses: Waste must be shredded and/or separated before entering the pyrolysis reactor to prevent blockage of the feed and transport systems which can substantially increase the cost for the installation and operation of the unit; pyrolytic oils/tars contain toxic and carcinogenic compounds; the pyrolysis products cannot be disposed without further treatment; the facilities for cleaning the gases and wastewater extremely cost-intensive; depending on the nature of the treated waste the solid residue contains 20–30% of the calorific value of the primary fuel (MSW) which requires final combustion in a solid fuel boiler or gasifier; char has a high content of heavy metals; back-up fuel supply is required for the operation, at least during start-up; long term operating experience from plants of significant scale is nil; nevertheless, the prospects for reactors of average temperature with the form of rotary drum or fluidized bed seem to be better (ref: Feasibility Study of Thermal Waste Treatment/Recovery Options in the Limerick/Clare/Kerry Region; and Moustakas and Loizidou 2010).

In summary, the general features of pyrolysis are:

- No oxygen is present during the process other than any oxygen contained in the fuel;
- Process temperatures vary from 400°C to 800°C;
- Process products are gas, liquid (pyrolytic oil), and char (material which is not completely oxidised);
- Lower temperatures with longer residence times tend to result in more char;
- Higher temperatures with short residence times (<1 s) tend to result in more liquid (up to 80%);
- Typical net calorific value (NCV) of the medium energy gas produced is 15 to 20 MJ/Nm³ at normal temperature and pressure.

4.2.4   Utilization of the process products

The syngas from the reactor can be: (a) combusted directly in a thermal oxidizer or boiler, making steam for power generation; the exhaust gases then pass through an emission control system that may include fabric filters or electrostatic precipitators for removal of particulate matter, wet or dry scrubbers for removal of acid gases, and activated carbon beds for removal of heavy metals, (b) Quench-cooled, cleaned in an emission control system, and then combusted in a reciprocating engine or gas turbine for power generation, and (c) Quench-cooled, cleaned in an emission control system, and then used for producing organic chemicals (CH2M Hill 2009). It can also be used to heat the pyrolytic reaction chamber or dry the feedstock entering the reaction chamber.

The pyrolytic oils can be used directly in fuel applications or can be refined for producing higher quality secondary products such as engine fuels, chemicals, and other products. The oil has a higher energy density than the raw waste.

The inorganic materials in the feedstock are removed as a bottom ash. It is usually combined with the unreacted char, and can be separated out for disposal or used in making construction materials. The solid char can be used as a solid fuel or as a char-oil, char-water slurry for fuel or can be used as carbon black or upgraded to activated carbon.

The utility of pyrolysis for secondary fuel production or materials recovery from waste depends on the presence of potential pollutants which could make the pyrolysis products difficult to use or even unusable.

4.2.5 Commercial scale pyrolysis plants

(A) *Commercial processes:* Several waste pyrolysis process in combination with combustion or gasification are developed and demonstrated (see Malkow 2004; BREF report 2001, GBB 2013).

Example 1: Pyrolysis unit added to a Power plant:
The ConTherm (pyrolysis) Plant was designed to integrate with a coal-fied Power plant located at Hamm-Unentrop, Germany. The plant consists of two lines of drum-type kilns with a planned annual municipal waste throughput of 50 kton each. The boiler unit in the Power plant would be supplied at full load with up to 10% of the pyrolysed fuels (coke and pyrolysed gas).

Shredded MSW and auto shredder residue (ASR) and upto 50% of used plastics are pyrolysed in indirectly heated rotary kilns in the absence of oxygen at 450°–550°C for 1 hour. The resulting products are coke, pyrolysis (product) gas, metals and inert materials. The metals entered with the feedstock, now present in their metal form, are recovered in a high purity state. The reutilization plant at the end of the rotary kilns separates the solid residue of the ConTherm plant into a coarse fraction (metals, inerts) and a fine fration (coke). 99% of the carbon is contained as coke in the fine fraction. After sifting, the coarse fraction is supplied to a wet ash remover, cooled down and separated into ferrous and non-ferrous metals in a reprocessing plant. The pyrolysis gas consists mainly of CO, H_2 and CH_4 and high-order carbohydrates. A cyclone ensures dedusting of the pyrolysis gas and the deposited dusts and carbon particles are added to the pyrolysis coke.

The Power plant's thermal output is 790 MW. Besides the regular fuels (coal, coke and petroleum coke), the pyrolysis products are also used as fuels (upto 10% substitution). The coke is first fed into the coal bunkers, ground together with the coal and then blown into the boiler with dust burners. The combustion of the pyrolysis

Table 4.3 Commercially active Pyrolysis Facilities using MSW (Source: SWANA 2011, Foth Infra. & Environ. 2013).

Location	Company	Operating since	MSW capacity
Toyohashi city, Japan	Mitsui Babcock	2002	*2 × 220 TPD
Aichi Prefecture			*77 TPD bulky waste facility
Mamm, Germany	Techtrade	2002	*353 TPD
Koga Seibu, Japan	Mitsui Babcock	2003	*2 × 143 TPD
Fukuoka Prefecture			*No bulky waste facility
Yame Seibu, Japan	Mitsui Babcock	2000	*2 × 121 TPD
Fukuoka Prefecture			*55 TPD bulky waste facility
Izumo, Japan	Thidde/Hitachi	2003	*7000 TPY
Nishi Iburi, Japan	Mitsui Babcock	2003	*2 × 115 TPD
Hokkaido Prefecture			*63 TPD bulky waste facility
Kokubu, Japan	Takuma	2003	*2 × 89 TPD
Kyouhoku, Japan	Mitsui Babcock	2003	*2 × 88 TPD
Prefecture			*No bulky waste facility
Ebetsu city, Japan	Mitsui Babcock	2002	*2 × 77 TPD
Hokkaido Prefecture			*38 TPD bulky waste facility
Oshima, Hokkaido Island, Japan	Takuma		*2 × 66 TPD
Burgau, Germany	Technip/Waste	1987	*40,000 TPY
Itoigawa, Japan	Thidde/Hitachi	2002	*25,000 TPY

products take place at about 1600°C when the organic agents are converted into CO_2 and water. The high ratio of sulphur-to-chlorine in the crude gas, and its fast cooling down to approximately 120°C prevents any new dioxins formation. All toxic agents that have not altered into their gaseous phase are bound into the melting chamber granulate together with the recycled airborne dust and the ground inert material.

Example 2: Pyrolysis – Combustion combination:
Three different processes are followed in the treatment of municipal waste:

(a) Smolder-burn-process pyrolyse in the drum-type kiln with subsequent high temperature incineration of pyrolyse gas and pyrolyse coke; (b) PyroMelt-process pyrolyse in the drum-type kiln resulting in condensation of the gaseous tars and oils; then subsequent high-temperature incineration of pyrolysis products (gas, pyrolysis oil and coke); (c) Duotherm-process pyrolyse on the grate with high-temperature incineration directly connected.

The solid residues from theses processes accumulate as melt granulate, which is advantageous for later application or disposal.

The waste gas cleaning systems for these processes are, in principle, same as those used in municipal waste incineration plants. The same residues and reaction products will accumulate and their nature and composition mainly depend on the chosen gas cleaning system. In contrast to municipal waste incineration, however, filter dusts are recycled into the melting chamber.

In these processes, dried or dehydrated sewage sludge may be co-treated.

Example 3: Pyrolysis-gasification combination processes are also developed. One of these is Thermoselect process discussed in the next section.
(B) *Commercial Projects*: There are several pyrolysis systems in Japan, Europe, Australia, Indonesia and other countries that use MSW as a feed (U.C. Riverside 2009).

MSW is typically used in combination with other wastes such as industrial waste, pet-coke, auto shedder residue, and medical waste. Japan may be considered as leader in applying pyrolysis technology to waste treatment. There are no known commercial MSW pyrolysis plants in North America. Twelve commercially-active pyrolysis facilities that use MSW are listed in Table 4.3 (SWANA 2011).

4.3   GASIFICATION

Gasification is centuries-old process, discovered in 1699 by Dean Clayton (Knoef 2005) and was applied during the nineteenth century for producing town gas. The first gas plant was established in 1812 in London. With the discovery of Franz Fischer and Hans Tropsch process (named after them) in 1923, it became possible to convert coal to liquid fuel. During World War II, German army needed to improve the use of the gasification process for fuel and chemical production. At the end of the war, the usefulness of this process was reduced but currently gained importance as the gasification of wastes appears as an interesting alternative for energy production (Fabry et al. 2013, Kokalj and Semac 2013).

Only in recent years the gasification process has been applied to the treatment of heterogeneous municipal solid waste (Lawrence 1998). MSW is composed of nearly 60% biomass derived components including wood, food scraps, yard trimmings, paper and plastics, which can be an advantage in terms of carbon content and high heating value. The elemental composition of MSW is in the range (weight %): carbon: 17–30, hydrogen: 1.5–3.4, oxygen: 8–23, water: 24–34, ashes: 18–43 and the average calorific value ranges from 5 to 10 MJ/kg (Cherednichenko et al. 2002). Studies on MSW gasification is limited compared to the investigations of the gasification of each component contained in MSW (Choy et al. 2004; Cheung et al. 2007; Jung et al. 2005).

Gasification differs from pyrolysis in the sense the process occurs at higher temperatures in the presence of a limited amount of oxygen. This oxygen reacts with carbon in the waste to produce combustible syngas having a significant heating value. The limited oxygen is made available through a gasification agent (air or CO_2 or steam) or directly. The syngas is composed mainly of CO and H_2 (nearly 85%) with low quantities of carbon dioxide, water, methane, hydrogen sulfide, ammonia, and under certain conditions, solid carbon, nitrogen, argon and some traces of tar. Nitrogen and argon arise from the use of air as the oxidant or as plasma gas. The syngas can be used as a fuel for heat/power production or further processed to obtain a wide variety of liquid fuels and chemicals (e.g., Ladwig et al. 2007; Ishikawa et al. 2008; Galeno et al. 2011, Khodakov et al. 2007, Oxtoby 2002, Young 2010).

The gasification process is influenced by several factors: gasification agent and temperature, reactor design, heating method, moisture content and particle size of feedstock and so on.

MSW is generally converted into RDF for its gasification because RDF has many benefits such as a higher heating value, more homogeneity in physical and chemical compositions, lower pollutant emissions, reduced excess air requirement during combustion, easier storage, handling, and transportation as explained in chapter 2. On average, 75%–85% of the weight of MSW is converted into RDF and approximately 80%–90% of the BTU value is retained. A typical treatment prior to the conversion to RDF can be found elsewhere (Jayarama Reddy 2011, 2014; Spliehoff 2010; Kordylewski 2005).

Industrial wastes are mostly homogenous – containing one type of waste with small variations in its composition. The processing of industrial waste for energy recycling does not require most of the steps followed regarding municipal waste (Kaminska-Pietrzak and Smolinski 2013).

In the gasification process, one can recover up to 80% of the chemical energy contained in the organic matter (Higman and Burgt 2003) compared to net electrical efficiencies of 18 to 22% theoretically achievable with a boiler-steam turbine system in an industrial-scale incineration plant (Yassin et al. 2009). A plasma gasifier associated with a gas turbine combined cycle (GCC) power plant can target up to 46.2% efficiency (Rutberg et al. 2011). Nonetheless, Waste combustion has a slightly higher thermal efficiency than gasification (Rezaiyan and Cheremisinoff 2005; Morris and Waldheim 1998; Quaak et al. 1999).

Gasification as applied to MSW is different than MSW incineration and has certain *advantages*:

(a)   The oxygen-deficient atmosphere in a gasifier does not provide the environment needed for dioxin and furans to form or reform as well as to produce large quantities of SO_x and NO_x (Fabry et al. 2013, AES 2004),

(b)   Dioxins need fine metal particulates in the exhaust to reform; syngas from gasification is typically cleaned of particulates before being used,

(c)   As the process requires just a fraction of the stoichiometric amount of oxygen necessary for combustion, the process gas volume is low requiring smaller and less expensive gas cleaning equipment. The lower gas volume also means a higher partial pressure of contaminants in the off-gas, which favours more complete adsorption and particulate capture,

(d)   In gasification facilities where syngas is used to produce fuels, chemicals and fertilizers, the syngas is quickly quenched, so that there is not sufficient residence time in the temperature range where dioxins or furans could re-form;

(e)   When the syngas is primarily used for producing heat, it can be cleaned as necessary before combustion; this cannot occur in incineration.

The syngas generated can be integrated with combined cycle turbines, reciprocating engines and, potentially, with fuel cells that convert fuel energy to electricity more efficiently than traditional steam boilers.

The ash produced from gasification is different from that produced from an incinerator. The incinerator ash is considered safe for use as alternative daily cover on landfills, but have concerns for use in commercial products.In high-temperature gasification, the ash actually flows from the gasifier in a molten form, where it is quench-cooled, forming a glassy, non-leachable slag that has more utility value. Some gasifiers are designed to recover melted metals in a separate stream, taking advantage of the ability of gasification technology to enhance recycling.

The gasification plants are modular. They are made up of small units which can be added to or taken away as waste streams or volumes change (e.g. with increased recycling). Therefore they are flexible and can operate at a smaller scale than mass-burn incinerators.

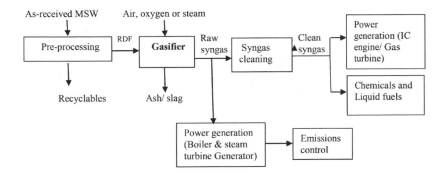

Figure 4.2 Flow diagram of conventional gasification of MSW.

On average, one ton of MSW in a gasification facility can produce up to 1000 kWh of electricity, nearly twice to that produced in an incinerator, proving it is a much more efficient method to utilize solid waste.

However, gasification and pyrolysis have some *disadvantages* too. Though the plants work with truly residual waste (after removing recycling materials such as metals, plastics and glass), they need a certain amount of plastics, paper, and food waste for effective gasification. A wide range of plastics cannot be recycled or cannot be recycled any further; such non-recycled plastics (NRP) offer an excellent feedstock for gasification because of their significant heating value, about 32.56 MJ/kg (Columbia University Earth Engineering Center). Disposal of ash and other by-products may be required, though some companies claim that their process makes this easier than for incineration ash. Secondly, during gasification, tars, heavy metals, halogens and alkaline compounds are released within the product gas and can cause environmental and operational problems. The tars are organic gases of high molecular weight that ruin reforming catalysts, sulfur removal systems, ceramic filters and increase the occurrence of slagging in boilers and on other metal and refractory surfaces. Alkalis can increase agglomeration in fluidized beds that are used in some gasification systems and also can ruin gas turbines during combustion. Heavy metals are toxic and accumulate if released into the environment. Halogens are corrosive and result in acid rain if emitted to the environment. The cost efficient and clean energy recovery from MSW gasification can be realized if the release and formation of these contaminants are effectively controlled (Klein 2002).

Two types of gasification, *Conventional gasification* and *Plasma arc gasification* are the established practices.

4.3.1 Conventional gasification

The flow chart of conventional gasification of MSW is shown schematically in Figure 4.2.

4.3.2 Chemical reactions in gasification

The feedstock is homogenized into smaller particles and inserted into the gasifier, followed by a controlled amount of air or oxygen. The feedstock passes through several

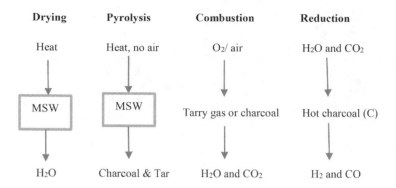

Figure 4.3 The four processes in gasification.

temperature zones where a sequence of reactions occurs before the syngas is produced and removed from the chamber. The temperatures in a gasifier typically range from 760 to 1500°C. The solid residue is removed from the bottom of the reaction chamber.

Typically, four distinct processes take place: drying, pyrolysis process in which tar and other volatiles are driven off, combustion (and cracking), and reduction. Though there is a considerable overlap of the processes, each can be assumed to occupy a separate zone where fundamentally different chemical and thermal reactions take place (Figure 4.3). The first two phases are common to combustion and pyrolysis processes.

Drying: The waste feedstock is heated below temperatures of 200°C when the moisture in it is released by evaporation. High moisture content fuel and/or poor handling of the moisture internally is one of the most common reasons for failure to produce clean gas.

i.e., wet feedstock + heat → dry feedstock + steam

Pyrolysis is heating of the feedstock in the absence of air/oxygen at high temperatures so as to break it down into charcoal (fixed carbon-to-carbon chains) and various tar gasses and liquids. It is essentially the process of charring (Rezaiyan and Cheremisinoff 2005).

dry feedstock + heat → char + volatiles components

The feedstock begins to rapidly decompose with heat once its temperature rises above around 240°C breaking down into a combination of solids, liquids and gasses. The solids that remain are *charcoal (char)*, and the gasses and liquids released are *tars*. The composition of products gas depends on the temperature and method of heat supply to the rector (Knoef and Ahrenfeldt 2005).

During *direct* heated gasification in the oxygen-deficient atmosphere, the light hydrogen-rich volatile hydrocarbons are released. Also, a gas product and solid carbon (char) and other components like tars, phenols and hydrocarbon gases are formed (Knoef and Ahrenfeldt 2005, Rezaiyan and Cheremisinoff 2005, Letellier 2008).

In *indirectly* heated gasification, steam is used as a heating source which acts as hydrogenation agent. Adding hydrogen to the system would accelerate the formation of hydrogen and CO. It means the steam acts as a catalyst for gasification reactions at relatively low temperatures. Further, methane formation is also promoted by the hydrogenation process within the gasifier at the same low temperature (Higman and Burgt 2008).

The amount of volatile components depends on the nature of the raw material and mainly consists of H_2O, H_2, N_2, O_2, CO_2, CO, CH_4, H_2S, NH_3 and C_2H_6. In addition, small amount of unsaturated hydrocarbons such as acetylenes, olefins, aromatics and aldehydes, tars and char are formed (Farokh Sahraei and Sara Akhlegi 2010).

Cracking is the process of breaking down large complex molecules such as tar into lighter gases through heat which is vital for the production of clean gas for use with an IC engine; otherwise, tar gases condense into sticky tar and rapidly fouls the valves of the engine. Cracking is also necessary to ensure proper combustion which occurs only when combustible gases mix thoroughly with oxygen. The high temperatures produced during combustion decompose the large tar molecules passing through the combustion zone.

Combustion is the only net exothermic process of all the phases; and all of the heat that drives drying, pyrolysis, and reduction comes either directly from combustion, or is recovered indirectly from combustion by heat exchange methods. Combustion can be fueled by either the tar gasses or char from pyrolysis. Different reactor types use one or the other or both. As will be seen later, in a downdraft gasifier, the tar gases from pyrolysis are burned to generate heat to run reduction, as well as the CO_2 and H_2O to reduce in reduction. The objective in combustion in a downdraft is to get good mixing and high temperatures so that all the tars are either burned or cracked so that they are not be present in the outgoing gas. The tar issue has to be solved only through cracking in the combustion zone.

Reduction is a process where oxygen atoms are stripped off combustion products of hydrocarbon molecules, to return the molecules to forms that can burn again. In fact, in most burning environments, these two processes operate concurrently in a dynamic equilibrium, moving back and forth. Reduction in a gasifier is accomplished by passing CO_2 or water vapor across a bed of red hot charcoal. The carbon in the hot charcoal is highly reactive with oxygen; due to its high oxygen affinity, it strips the oxygen off water vapor and carbon dioxide, and redistributes it to as many single bond sites as possible. The oxygen is more attracted to the bond site on the carbon than to itself, thus no free oxygen can survive in its usual diatomic form. All the available oxygen would bond to available C sites as individual O until all the oxygen is gone. After the redistribution of all the available oxygen as single atoms, reduction stops. CO_2 is thus reduced by carbon to produce two CO molecules, and H_2O is reduced to produce H_2 and CO. Both H_2 and CO comprise the combustible product gas (syngas).

Thus, the waste conversion into syngas involves complex chemical reactions (de Souza-Santos 2004). The main chemical reactions of gasification occurring after the pyrolysis of the wastes are shown in Table 4.4 (Zhang et al. 2010; Arena 2012). Heterogeneous reactions take place in gas-solid phase while the homogeneous reactions occur in gas-gas phase. The reactions represented by eqns. 4, 6, 7, 8, 11 and 12 are most basic and important (Krigmont 1999), and are reversible, their rates depending on the temperature, pressure and concentration of oxygen in the reactor.

The homogeneous reactions (eqns. 9 to12) are almost instantaneous in high temperature conditions in contrast to heterogeneous reactions (eqns. 1 to 8). Among the

Table 4.4   Main chemical reactions of gasification.

Eqn	Reaction	Chemical reaction	Enthalpy_H **
I	$C_nH_mO_k$ partial oxidation	$C_nH_m + (n/2)\,O_2 \rightarrow (m/2)\,H_2 + n\,CO$	Exothermic
2	Steam reforming	$C_nH_m + n\,H_2O \rightarrow (n+m/2)\,H_2 + n\,CO$	Endothermic
3	Dry reforming	$C_nH_m + n\,CO_2 \rightarrow (m/2)\,H_2 + 2n\,CO$	Endothermic
4*	Carbon oxidation	$C + O_2 \rightarrow CO_2$	-393.65 kJ/mol
5	Carbon partial oxidation	$C + \frac{1}{2}\,O_2 \rightarrow CO$	-110.56 kJ/mol
6*	Water-gas reaction	$C + H_2O \rightarrow CO + H_2$	$+131.2$ kJ/mol
7*	Boudouard reaction	$C + CO_2 \rightarrow 2CO$	$+172.52$ kJ/mol
8*	Hydrogasification	$C + 2\,H_2 \rightarrow CH_4$	-74.87 kJ/mol
9	Carbon monoxide oxidation	$CO + \frac{1}{2}\,O_2 \rightarrow CO_2$	-283.01 kJ/mol
10	Hydrogen oxidation	$H_2 + \frac{1}{2}\,O_2 \rightarrow H_2O$	-241.09 kJ/mol
11*	Water-gas shift reaction	$CO + H_2O \rightarrow CO_2 + H_2$	-41.18 kJ/mol
12*	Methanation	$CO + 3\,H_2 \rightarrow CH_4 + H_2O$	-206.23 kJ/mol

*Most basic reactions and reversible.
**T $= 298°$K, P $= 1.013$ 105 Pa, carbon as solid and water in vapor form.
(Source: Fabry et al. 2013).

large number of gasification reactions, three of them are independent reactions: Water-gas reaction (eqn. 6), Boudouard reaction (eqn. 7) and Hydrogasification (eqn. 8). In the gas phase, these reactions can be reduced to only two: Water-gas shift reaction (eqn. 11) which is the combination of the reactions, eqns. 6 and 7, and methanation, eqn. 12, which is the combination of the reactions, eqns. 6 and 8.

It is important to note that all these gasification reactions, except the oxidation ones, are equilibrium reactions. The final composition of the raw syngas is determined by reaction rates and also by the effect of catalysts which is important for tar decomposition in the reactor, rather than by the thermodynamic equilibrium (Arena 2012).

4.3.3   Key factors for gasification of waste

The factors that are very vital and significantly impact the gasification performance are briefly outlined:

(a) Gasification agent: The gasification agent such as air, oxygen, steam or carbon dioxide which promotes the gasification process has a significant effect on the system performance and product gas composition (Bridgwater 2003, Colpan et al. 2010). With air as the gasification agent, there was rather high amount of nitrogen in product gas, about 47.5% by volume dry basis and low heating value, around 5.5 MJ/Nm³. The main products are CO, CO_2, H_2, CH_4, N_2 and tars. Using oxygen, the product syngas contains no nitrogen or low amount of nitrogen, around 22.5% by volume dry basis depending on the purity of the oxygen. The other products are same as above. The cost of providing and using oxygen is compensated by a better quality of syngas. Gasification of biomass and MSW with pure steam can improve the heating value of the syngas from ~15 MJ/Nm³ to 30 MJ/Nm³ (Colpan et al. 2010; Bi and Liu 2010) by reducing the nitrogen content in the syngas depending on purity of the steam and ratio of steam to oxygen (Bi and Liu, 2010). The process has two stages with a primary reactor producing gas and char, and a second reactor for char combustion to reheat

Table 4.5 Effect of temperature on gas composition.

Temp. °C	H_2 mol%	CO mol%	CO_2 mol%	CH_4 mol%	HHV (MJ/m^3)
500	5.56	33.5	44.8	17.50	12.3
650	16.6	30.5	31.8	11.0	15.8
900	32.5	33.5	18.3	4.50	15.1

(Source: Farokh Sahraei & Sara Akhlaghi 2011).

sand which is recirculated. The gas heating value is maximised due to a higher methane and higher hydrocarbon gas content, but at the expense of lower overall efficiency due to loss of carbon in the second reactor (TwE 2014). Steam as gasification agent considerably increased the yield of H_2, CO and CH_4 (He et al. 2009; Colpan et al. 2010). The required steam can be generated through a boiler or a heat recovery steam generator (HRSG) included in the plant design.

Using high-temperature gasification agent, the quality of the product gas becomes less sensitive to variations in the particle size, heating value, and moisture content of the MSW (Anna et al. 2006). The features of 'high-temperature agent gasification' are more significant if high-temperature steam is used as the gasification agent (Nipathummakkul et al. 2010, Young et al. 2011).

Temperature: The reactor temperature is a key parameter of gasification process. The higher temperatures promote the reaction rate, the product gas yield, hydrogen content and the complete conversion of char. The lower temperatures promote methane formation, char tar, and volatile content (Higman and Burgt 2008; He et al. 2010). At increasing temperatures less methane and less CO_2 are formed (Castaldi 2008).

As seen in Table 4.5, the CO content decreased from 500°C to 650°C and then increased again at 900°C. The maximum concentration of CO and H_2 can be obtained at temperature of about 900°C. Also, gradually increase of temperature may lead to increase of H_2 and high heating value of product gas at equilibrium condition for the gasification. Thus, the gasification temperature can influence the concentration of the desired product, CH_4, H_2, or CO (Higman and Burgt 2008; Wang and Kinoshita 1993). However, it must be noted that the reaction kinetic, fuel composition and eventual catalytic material also have a significant effect when the optimum operation conditions are determined (Gomez-Barea 2010).

Pressure: The operating pressure of a gasifier is selected according to process requirement and products applications (Higman and Burgt 2003). For instance, for ammonia production, the pressure of the gas should be in range, 130–180 bars; and for use in a gas turbine for power production the pressure needs to be around 20 bars. The effect of pressure on the components of the product gas at 1000°C is shown graphically in Higman and Burgt (2003). The studies reveal that increasing pressure makes a gradual increase in CH_4 and CO_2, and slight decrease in CO and H_2 contents in product gas (Castaldi 2008).

In industrial applications of syngas where high pressures are required, it is desirable to pressurize the gasifier to reduce energy consumption and equipment size, and also for decreasing collection of the ashes inside the reactor (Knoef and Ahrenfeldt 2005).

Majority of gasifiers operate generally in the sub-atmospheric pressure range so that, in case of leakage, the outside air can enter into the reactor preventing the

Table 4.6   Effect of heating method on heating value of product gas.

Producer gas	Gasification agent	HHV of Product gas (MJ/Nm³)
Direct heated gasification	Air	4–6
Pure oxidation gasification	Oxygen	10–12
Indirect heated gasification	Steam	15–20

(Source: Higman & Burgt 2008, 2003).

potential hazardous gases to escape. Further, for high pressure gasification, the cost of the reactor and the whole system would increase because the reactor materials and gasifier joints must be chosen to withstand such pressures, in addition to looking into isolation issues and feedstock feeding equipment.

Moisture: The moisture level impacts the reaction temperatures and composition of syngas as well as the energy balance of the process in several ways. The allowable range of moisture content is about 10–15% (Rezaiyan and Cheremisinoff 2005).

Heating rate: The effect of the heating rate can be seen on volatiles separation during the pyrolysis and the devolatilization step. A higher heating rate rapidly releases the volatiles and significantly increases the porous nature and reaction rate of char in the steam gasification of biomass. On the other hand, low heating rate allows char particles to react with volatiles (Higman and Burgt 2008). The heating rate does not influence the elemental composition of char (Fushimi et al. 2003).

Heating method: The gasification process involves a series of endothermic and exothermic reactions. In the direct-heated gasifier, the necessary heat is provided by exothermic reactions through a partial oxidation of the feedstock inside the reactor; and in the indirect heated gasifier, the required heat is provided by an external source. The heating value of product gas in the former gasifier is greater than the later as shown in Table 4.6 (Higman and Burgt, 2008, 2003; Belgiorno 2003).

Feedstock heating value: The heating value in general terms refers to the amount of heat released from the combustion reaction expressed in weight unit for solid fuel (MJ/kg), and volume unit for gas fuel (MJ/Nm³). Heating value of feedstock depends on the moisture content and combustible organic material (De Filippis et al. 2004). Among different feedstock, the highest value is for coal; and for the biomass-type fuels it is in the range of 14–20 MJ/kg (Higman and Burgt 2008, 2003; Van Wylen et al. 1994).

Waste particle size and preparation steps: The density, size and shape of the waste particles influence the heat transfer within the gasifier bed. For instance, an entrained flow gasifier requires the particle size of the feed in the range of a few hundred μm; for a fluidized bed reactor, in the range of a few mm; and larger particles in a fixed-bed gasifier (Rezaiyan and Cheremisinoff 2005). The large-sized feed particles are left untreated in the process (Knoef and Ahrenfeldt 2005; Bridgwater 1995).

In summary:

(a) CH_4 formation decreases with increasing temperature and increases with increasing pressures,

(b)   CO and H_2 formation increases with increasing temperature and reducing pressures,

(c) CO_2 concentration increases with increasing pressures and decreases sharply with increasing temperatures,

(d)   Maximum concentration of H_2 and CO can be obtained at atmospheric pressure and temperatures of 800 to 1000°C,

(e)   Reducing oxygen-to-steam ratio of reactant gases (or reactor inlet streams) increases H_2 and CH_4 formation, and increasing the oxygen-to-steam ratio enhances CO and CO_2 formation,

(f)   Gasifier temperature and pressure can be controlled to maximize the concentration of desired product, CH_4 or H_2 and CO.

Ashes and pollutants

The composition and melting point of the ash have great impact on ash behavior particularly at high temperatures, and on the accumulation rate in the reactor. Moreover the ash quantity can influence the ash discharge system and the gasifier type (Gomez-Barea 2009).

Pollutants can be classified into tar, particles and heavy metals. Tar refers to condensable organic or inorganic compounds present in raw product gas, and is a limiting factor in the gasification process. Tars make fouling or inactivate catalytic filters (Gomez-Barea 2010). The mineral substances in the feed, the unreacted solid materials and part of fuel entrained with gas products result as pollutants. These pollutants can be removed by cyclones, filters or scrubbers. MSW feedstock has quite a low level of heavy metals. If they are present in high amount, they must be removed through, for example, filtration (Letellier 2008). The raw product gas contains both gas and particle impurities that may have negative effects on both catalysts and hot-gas filters (Eva Gustafsson and Mehri Sanati 2007).

Gas utilization

To utilize the syngas for deriving products, gas cooling and cleaning are required. The quality of syngas depends on the key factors described so far. Cleaning process involves removing unwanted components including particulates, alkali, tars, sulfur, and ammonia. A gas cleaning system may consist of cyclone separation, gas cooling system, low temperature gas cleaning, high temperature gas cleaning, acid removal, sulfur recovery, CO_2 removal, and gas reforming.

The product gas can be upgraded to hydrogen-rich syngas. The syngas can be further converted to liquid by Fischer Tropsch synthesis or gaseous fuels and chemicals, including fuels such as methanol, dimethyl ether (DME), and synthetic diesel.

4.3.4   Gasifier configurations

There is considerable variety in the specific designs of gasifiers. Many of these variations are derived from specific application needs regarding process scale, gas quality, feedstock feeding, and ash management. However, from a flow dynamics perspective, there are four general types of gasifier. The basic design of each type is built around the reaction chamber with insertion of feedstock, but each has a different heating mechanism, air entry and syngas removal location. Each general gasifier type has unique benefits and drawbacks depending not only on the feedstock but the end product as

Table 4.7 Different types of gasification reactor configurations.

Gasification reactor type	Mode of contact
Fixed Bed:	
Downdraft	Solids and Gas both move downwards
Updraft	Solids move downwards, and Gas moves upwards
Cross-draft	Solids move downwards; Gas moves at right angles
Variants	Stirred bed; Two stage gasifier
Fluidized Bed:	
Bubbling bed	Low gas velocity; inert solids stay in reactor
Circulating bed	Higher gas velocities; inert solids elutriated, separated and recirculated
Entrained bed	No inert solid; highest gas velocity of lean phase systems
Twin reactor	1st stage: steam gasification; 2nd stage: char combustion
Moving bed:	
Variants	Mechanical transport of solid, usually horizontal; typically used for pyrolysis i.e., lower temperature processes
	multiple hearth, horizontal moving bed, sloping hearth, screw/augur kiln
Other:	
Rotary kiln	Good solid-gas contact
Cyclonic reactor	High particle velocity and turbulence to effect high reaction rates

(Source: WSP 2013).

well (Fabry et al. 2013). Detailed descriptions of these configurations are given in several publications, for example, WSP (2013), Arena (2012), E4Tech (2009), PNNL (2008), Ladwig et al. (2007), Chopra and Jain (2007), Bridgwater (2003), Klein and Themelis (2003) and so on.

The main reactor configurations are 'fixed bed (downdraft and updraft)', 'fluidized bed (bubbling and circulating)', 'entrained flow', 'rotary kiln', and 'Moving grate' (Table 4.7). Of these, the commonly used configurations are Fixed-bed (updraft and downdraft), fluidized bed, and entrained flow types (e.g. WSP 2013, Jayarama Reddy 2013, Arena 2012, Reed and Gaur 2001). Other gasifier types, including plasma gasifiers do not rely on a different gasifier structure or arrangement of air inlets and syngas outlets but rather on type of heat source used.

4.3.4.1   Fixed bed gasifiers

The fixed bed gasifiers are relatively simple in design, and most of them are used to treat biomass (e.g., Lv et al. 2007; Saravanakumar et al. 2007; Friberg and Blasiak 2002; Dogru et al. 2002; Niu et al. 2001). The waste is fed at the top of the reactor, and the ash and unreacted char removed from the bottom. As the feedstock does not move freely in the reactor, they are called fixed bed. Instead it moves slowly downward through a fixed zone supported by a grate at the bottom of the gasifier. Fixed bed gasifiers are relatively small in size and are directly heated using air or purified oxygen or steam as oxidizing/gasification agent.

As mentioned already, gasification with air results in a nitrogen-rich, low-Btu fuel gas. Gasification with pure oxygen results in a higher quality mixture of CO and hydrogen and virtually no nitrogen. Gasification with steam is generally called 'reforming' and results in a hydrogen- and CO_2-rich syngas. After cleaning from contaminants, the

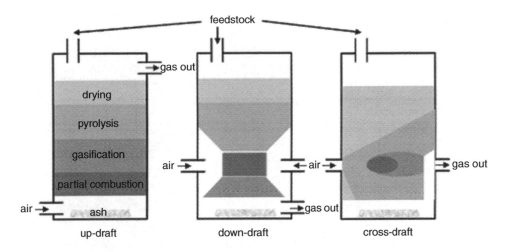

Figure 4.4 Simplified Schematic of an Up-draft, Down-draft and Cross-draft gasifiers (Source: GBB 2013; reproduced with the permission of GBB).

syngas can be combusted in a boiler, producing steam for power generation (Jenkins 2007).

There are three fixed bed gasifier designs: *up-draft, down-draft and cross-draft.*

An Up-draft (counter-current flow gasifying agent and feedstock) has stacked zones clearly defined to dry, pyrolyze, gasify, and partial combust the feedstock. In this design, the wastes are fed by the top of the reactor. The gasification agent is introduced from the bottom of the reactor. The reacted gases exit the top of the reactor. The gasification reaction takes place in the bottom of the reactor between the down-coming waste and the ascending gas. The rise of the hot gas starts waste pyrolysis at lower temperatures and dries it. The tars that are carried in the crude gas, a disadvantage with this type of reactor configuration, have levels between 10% and 20% by weight (Ciferno and Marano 2002), which makes them difficult to clean for electricity generation. Char produced from slow pyrolysis of the waste continues downward to a grate where oxidizing agent is introduced combusting the char. Heat from this combustion process drives the endothermic drying and pyrolysis reactions. Ash and unreacted char drop through the grate and are removed from the gasifier. Because the gas leaves the gasifier at relatively low temperatures, the process has a high thermal efficiency. As a result, MSW containing 50% moisture can be gasified without any pre-drying of the waste. As it can handle high ash content with minimal processing it is *highly suited* to MSW. A simplified schematic of an up-draft gasifier is given in Fig. 4.4.

In *down-draft gasifiers* (co-current flow oxidizing agent and feedstock), the waste is added at a mid or top part of the reactor which moves downward, slowly drying and pyrolyzing. The moisture, pyrolysis gases, and tars produced as the waste slowly pyrolyzes pass downward to the bottom of the gasifer vessel passing through a zone where the gasification agent is introduced to the gasifier. The gasification agent combusts the pyrolysis gases, tars, and a portion of the char producing a high temperature zone (approximately 1800°F), where steam and CO_2 in the gas stream can react with

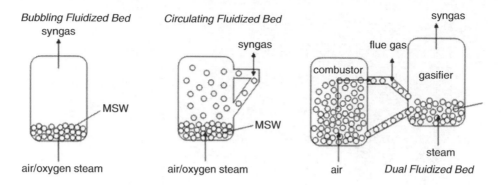

Figure 4.5a Simplified schematics of BFB gasifier, CFB gasifier and Dual FB gasifier.

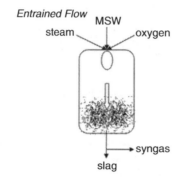

Figure 4.5b Schematic of Entrained flow gasifier.

the remaining char producing additional hydrogen and carbon monoxide (Cirferno and Marano 2002). The gas stream while leaving at the bottom of the reactor must go through the ash in the form of char which reduces the amount of tars in the syngas. The ash and unreacted char fall through the grate and are removed.

The tar levels in this reactor configuration are very low, around 0.1%, as the major part of tars is burned to supply the energy for the pyrolysis/gasification reactions of the wastes. This reactor configuration is *particularly suitable* for the production of clean gas requiring low post-treatment for their use in electricity production using gas turbines, at least in smaller-scale units, 8 kWe–500 kWe. However, the operation generally requires a long residence time, 1 to 3 hours (Chopra and Jain 2007). But it has the disadvantage of having low energy efficiency but with low tar concentrations (Knoef 2005). A simple schematic of a down-draft gasifier is given in Figure 4.4.

In a *Crossdraft* gasifier, the air inlet and the gas outlet are on the opposite sides in the middle of the reactor (Figure 4.4). This type of gasifiers are *less common* as they produce high temperature syngas at a high velocity that does not have as efficient carbon dioxide reduction as other gasifier types. The types of feedstocks for these systems are limited by the system design to low ash fuels, such as wood, petroleum coke, and charcoal. Crossdraft gasifiers have several advantages, including high carbon monoxide, low hydrogen and low methane syngas content when used on *dry fuels*, and a fast startup time desirable for some applications (GBB 2013).

In summary, fixed bed gasifier designs are simple, low-cost and easy to operate and are therefore useful for small- and medium-scale power and thermal energy use. However, it is difficult to keep operating temperatures at constant levels and to ensure adequate gas mixing in the reaction zone. As a result, gas yields can be unpredictable and are not optimal for large-scale power purposes (i.e. over 1 MW) (Klein, 2002). Therefore, larger capacity gasifiers are preferable for treatment of MSW because they allow for variable fuel feed, uniform process temperatures, good interaction between gases and solids, and high levels of carbon conversion (Climatetechwiki website).

Hybrid fixed bed type: There are hybrid fixed bed gasifier designs that possess elements of updraft and downdraft gasifiers. Baffles and alternative points of air introduction create combustion zones that may not be fully mixed with all gases and tars evolved in cooler parts of the gasifier as would occur in a downdraft gasifier. However, those gases and tars do experience much higher temperatures than are experienced in an updraft gasifer, providing opportunities for tar cracking (PNNL 2008).

4.3.4.2 Fluidized bed gasifiers

Fluidized beds offer an attractive plan and the best design for the large-scale gasification of MSW. Fluidized bed gasifiers are one type of updraft gasifier. In a fluidized bed boiler, inert material (sand or limestone) and solid feedstock are fluidized by a stream of air passed upward through the bed. The air acts as the fluidizing medium and also provides the oxidant for combustion and tar cracking. The fluidized bed behaves like a boiling liquid and has some of the physical characteristics of a fluid (suspension). The feedstock is introduced either from the top or from the side and mixed with the hot sand causing rapid pyrolysis to produce mostly gas and a small quantity of tar and char.

Fluidized-beds have the advantage of extremely good mixing and high heat transfer, resulting in very uniform bed conditions and efficient reactions. In this configuration, there is a mixture of the two phenomena identified in the updraft and downdraft reactors. Thus, the tar rate is at an intermediate level between the updraft and downdraft reactors, between 1% and 5%. These tars must be reduced to much lower levels for syngas applications, either by separation or by conversion into gas. The disadvantage of this configuration is the high proportion of particulates (tars) in the exhaust gas that requires high gas treatment and has low mass and energy yields (Knoef 2005; Fabry et al. 2013).

With fluidized bed gasifiers, there are limitations on the physical properties of the feedstock because the solids must be fluidizable or removable from the bottom of the vessel. As MSW contain a wide variety of materials, those items that cannot be readily gasified or that would harm the gasification equipment need to be sorted and removed. Further, with MSW, size reduction and screening to achieve particles with similar aerodynamic properties with respect to fluidization is essential. Also, the gasification system may have to be so designed as to handle a variety of different materials whose gasification rates may differ. Therefore, converting MSW in the form of RDF would be particularly suitable *as a feedstock*.

Two main types of fluidized beds are in practice (Figure 4.5a): In a *Bubbling Fluidized Bed (BFB)*, the gas velocity must be high enough to lift the solid particles

Table 4.8 Thermal capacity of different gasifier
           designs.

Gasifier design	Fuel capacity
Downdraft	1 kW –1 MW
Updraft	1.1 MW –12 MW
BFB	1 MW –50 MW
CFB	10 MW –200 MW

(Source: Morris 1998).

comprising the bed material, thus expanding the bed and causing it to bubble like a liquid. A bubbling fluidized bed reactor typically has a cylindrical or rectangular chamber designed so that the contact between the gas and solids facilitates drying and size reduction (attrition). As waste is introduced into the bed, most of the organics vaporize pyrolytically and are partially combusted in the bed. The desired operating temperatures typically range from 900° to 1000°C. The main concern with the bubbling-bed boilers is bed agglomeration. High alkaline content fuels cause particles in the bed to agglomerate, eventually defluidizing the system.

A *circulating fluidized bed (CFB)* is differentiated from a bubbling fluid bed in that there is no distinct separation between the dense solids zone and the dilute solids zone. The capacity to process different feedstock with varying compositions and moisture contents is a major advantage in such systems.

Generally, gasification technology is selected on the basis of available fuel quality, capacity range, and gas quality conditions. Table 4.8 shows the thermal capacity ranges for the main gasifier designs.

Entrained flow gasifiers (sometimes referred as circulating bed gasifiers) are similar to fluidized bed gasifiers except that the gases introduced into the bottom of the gasifier vessel are at a much higher velocity causing the fluidized medium and the feedstock to become entrained and carried out the top of the gasifier. Solids residence times in the gasifier range from 1 to 10 seconds and the temperatures are higher (1650°F–2550°F). The solids from the product gas are removed, and either returned to the gasifier or sent to a separate combustor where the char is combusted.

An entrained bed gasifier may be directly fired with air or oxygen or indirectly heated (using the separate combustor to heat the sand). The performance of entrained flow gasifiers are similar to those for fluidized bed type except the throughputs can be greater, but the solids properties are usually more stringent. The high velocities of the entrained solids may also accelerate equipment erosion, compared to a fluidized bed gasifier.

4.3.4.3   Slagging gasification

Most of the gasification processes currently in operation in Japan are designed to melt the inorganics present in the waste because the main objectives are waste volume minimization and maximized recycling. The resulting slag is inert and can be easily recyclable as a construction material. To slag the inorganic ash, high temperatures are usually required and this is facilitated by the utilization of oxygen rather than air for gasification and/or the use of plasma processes to provide the necessary input of heat energy. The production of oxygen is energy intensive and costly. Hence, in Japan, the Pressure Swing Adsorption systems are being used which are much more cost effective compared to cryogenic air separation systems.

Majority of commercial- scale operating facilities world-wide, primarily in Japan, employ slagging gasification, for example, SVZ Schwarze Pumpe process, EBARA process and so on.

4.3.5 Performance criteria

Different criteria are followed for defining performance of the different gasification processes. Normally, parameters such as 'energy efficiency (also called 'cold gas efficiency')', 'hydrogen yield rate' and 'CO yield rate' are defined (Fabry et al. 2013).

Cold gas efficiency is the energy produced by syngas combustion divided by the energy produced by direct combustion of product incremented by the added energy (electric or fuel) for allothermal (indirectly heated) processes. This efficiency does not take into account the steam consumption and electricity (used for pure oxygen production), or heat recovery by cooling syngas (steam). Fuel gas production is the flow of the gas mixture produced by gasification per kilogram of product treated in the reactor. When air is used as gasification/oxidizing agent in the reactor, the following equation can be used.

$$\text{Fuel gas production } (\text{Nm}^3 \cdot \text{kg}^{-1})$$
$$= \frac{\text{Air flow rate } (\text{Nm}^3 \cdot \text{s}^{-1}) \times 0.79}{[1 - (CO + CO_2 + H_2 + CH_4 + C_2H_2)/100] \times \text{feeding rate } (\text{kg} \cdot \text{s}^{-1})}$$

In the above equation, fuel gas production is function of the ratio of the nitrogen at the beginning of the process to the nitrogen in the mixture produced. In this particular case, which cannot be applied for all gasification situations, it is assumed that the conversion is total (no oxygen in the crude gas) and the only gases produced during gasification are CO, CO_2, H_2, CH_4 and C_2H_2. Further, the feedstock is composed of C and H only (no chemical species like S, Cl . . .).

Cold gas efficiency (η) is expressed as (Fabry et al. 2013):

$$\eta = \frac{\text{LHV of cold gas } (\text{kJ} \cdot \text{Nm}^{-3}) \times \text{fuel gas production } (\text{Nm}^3 \cdot \text{kg}^{-1})}{\text{LHV of waste treated } (\text{kJ} \cdot \text{kg}^{-1}) + \text{allothermal power } (\text{kW})/\text{waste flow rate } (\text{kg} \cdot \text{s}^{-1})}$$

In the case of waste gasification by thermal plasma, the electric energy consumed to create the plasma has to be taken into account. If the electric energy comes from the electrical power generated by the process, the allothermal power is equal to the electric energy consumed to create the plasma. If the electric energy comes from a primary thermal power plant, Allothermal Power = P_{plasma}(electrical)/Conversion efficiency of the thermodynamic cycle – Carnot cycle).

Generally, the conversion efficiency is between 30% and 40% for a single cycle steam power plant and can be up to 60% for a Combined Cycle Gas Turbine (CCGT) power plant.

H₂ and CO yields

H_2 yield is defined as the ratio of the mass of hydrogen in the syngas produced to the mass of hydrogen introduced. For the CO yield, it is the ratio of the mass of carbon atoms in the CO produced to the mass of the carbon atoms injected. Petitpas et al. (2007) expressed these ratios as:

H_2 yield = (H atoms in the syngas)/(H atoms injected)

CO yield = (C atoms in the formed CO)/(C atoms injected)

It is to be noted the H_2 rate in the crude syngas is strongly linked to the oxidizing agent injected and/or the moisture content in the waste treated. As H_2 yield is not representative of the conversion rate of the processes, only the CO yield can be used to provide proper information on the mass balance and on the performances of the gasification processes.

4.3.6 Tar content in syngas

The presence of tar in syngas is the most challenging issue in industrial gasification processes. Tar content has important implications in the design and the operation of gasifiers to ensure adequate control of reaction conditions. These tar constituents can be used as indicators of overall reactor performance and design (Brage et al. 1997). The nature of tar and its quantity mainly depend on the processing conditions, the applied technology and the nature of the waste.

Tars are complex mixture of condensable hydrocarbons or organic compounds having a molecular weight higher than benzene C_6H_6. This definition was introduced by the tar protocol measurement at the IEA Gasification Task meeting at Brussels in 1998 (Neeft et al. 1999).

The tar rate $(g.m^{-3})$ is representative of the quantity of tars mixed with the syngas after gasification of the organic material.

Depending on application desired, the syngas has to be cleaned of tar content. Mostly, tar removal was done through physical processes and tar conversion through thermo-chemical and catalytic processes (thermal, steam, partially oxidative, catalytic and/or plasma processes). The choice of the cleaning process depends on the specific application of the syngas.

Considerable volume of material on the tar reduction, conversion and/or destruction in waste gasification processes has been published (Milne 1998).

Tar levels from gasifiers

The tar rates from the three main types of gasifiers are given in Table 4.9, showing a wide range of values and the relative orders of magnitude (Milne 1998).

The updraft gasifiers appear to be the 'dirtiest' with large range of tar rates (average value, $100 \, g.Nm^{-3}$), and the 'downdraft' the 'cleanest' with the lowest range (average value, $1 \, g.Nm^{-3}$). The 'fluidized beds' have intermediate ranges with an average value, $10 \, g.Nm^{-3.}$

The choice of the gasifier technology depends on several fuel characteristics such as the particle size, the morphology, the moisture content, the ash content, the ash melting

Table 4.9 Tar and solid particles rates in the gasification raw-gas in different reactor configuration.

Reactor	Tar rates (g.Nm^{-3})			Solid particles rates (g.Nm^{-3})		
	Min	Max	R.R	Min	Max	R.R
Updraft	1	150	20–100	0.1	3	0.1–1
Downdraft	0.04	6	0.1–1.2	0.01	10	0.1–0.2
Fluidized bed	<0.1	23	1–15	1	100	2–20

R.R.: Representative Range in which are most of the processes studied.

point and the bulk density in addition to the temperature profile in the gasifier, the heat exchange, the residence time, the conversion efficiency and the process flexibility.

The limitations and the kinds of materials used as feedstock in gasifiers were discussed in several publications, for example, in an excellent review by Arena (2012).

4.4 PLASMA GASIFICATION

Conventional gasification methods present some limitations in material yield, syngas purity, energy efficiency, dynamic response, compactness and flexibility. These might be overcome by using thermal plasma for heat supply (Boulos 1996). When waste is introduced into the plasma field, the intense heat breaks the waste products' molecules into simpler compounds. These gaseous products are then scrubbed to remove contaminants, and burned or used directly in a gas turbine to produce electricity. In addition to this product gas, a glassy slag residue from the inorganic component of the waste also results.

The plasma arc torch produces a very high temperatue plasma gas (up to 15,000°C). The reaction chamber heated by the plasma will reach around 2000°C. The slag is typically around 1700°C when discharged from the pyrolysis chamber. The gasification reactions discussed earlier hold good for the plasma arc gasification also.

Thermal plasma process offers certain merits over the conventional methods: (i) high energy density (up to 100 MW/m) and reaction temperatures (up to 7000°C or more) and the correspondingly fast reaction times offer the potential for a large throughput with a small furnace, (ii) high heat flux densities at the furnace boundaries lead to fast attainment of steady state conditions allowing rapid start-up and shut-down times compared with other thermal processes such as incineration, (iii) only a small amount of oxidant is necessary to generate syngas; consequently, the gas volume produced is much smaller than with conventional combustion processes enabling an easy and less expensive approach to handle, (iv) capability to safely dispose hazardous wastes including medical waste, asbestos, munitions etc., (v) acid gases are readily neutralized, and (vi) production of inert vitrified solid after cooling from inorganic components in the waste which is highly resistant to leaching (Byun et al. 2013; Wilson and Wilson 2013; Circeo 2008).

Plasma gasification at high temperatures (up to 7000°C) was preferred for disposal of hazardous wastes by industries. Last few years, this technology was regarded as a

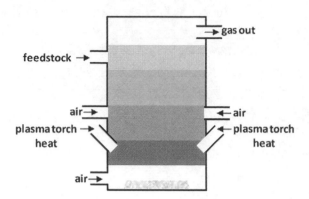

Figure 4.6 Plasma Arc gasification system (Source: GBB 2013; reproduced with the permission).

viable alternative to treat extremely toxic wastes such as APC residues, radioactive and medical wastes (e.g., Poiroux et al. 1996, Jimbo 1996, Krasovskaya et al. 1997, Chu et al. 1998, Tzeng et al. 1998, Inaba et al. 1999, Katou et al. 2001, Rutberg et al. 2002, Nema et al. 2002, IAEA 2006, Moustakas et al. 2005, Heberlein and Murphy 2008).

4.4.1   Plasma arc gasifier

Two different plasma gasification configurations are available: plasma assisted gasification and plasma coupled with traditional thermal gasification. The first type has the plasma torch/torches in the gasification reactor where the heat generated breaks apart the chemical bonds in the feedstock and forms gas. Inorganic rejected materials are collected at the bottom of the gasification chamber, as a glass-like inert material potentially suitable for construction or other aggregate applications.

Most plasma arc gasifiers are arranged similar to an updraft system as shown in Figure 4.6 (GBB 2013), where feedstock is inserted near the top of the chamber, the oxidizing agent in the middle or bottom of the chamber, and the exit of the syngas at the top of the reactor above the waste bed. The feedstock moves downward and into the intense heating zones created by the plasma torches. This type of system helps to prevent tar formation, as the syngas remains at a very high temperature (>1000°C) as it exits the chamber.

As extremely high temperatures are possible, extensive waste sorting is not required because all compoents of the waste would eventually be converted in the reactor as gas or as a molten slag. However, the operation of small specialty facilities and the demonstration plants had shown that the consistency of the municipal solid waste has a direct impact on the performance of a plasma facility. Waste streams that include large amounts of inorganic materials such as poorly sorted construction waste, metals, and glass, result in increased slag and decreased syngas production. The heat energy that is required to melt these inorganics is lost since the molten slag does not contribute to syngas production.

The selection of an optimal gasifier design for a particular application depends on variables such as the size, moisture content, and calorific value of the feedstock and the desired product type and quality.

A typical plasma system for the treatment of solid wastes consists of (a) plasma furnace, with a metal and the slag collection at the bottom that periodically tapped and cast into some usable form and power supply, cooling water supplies, gas supplies, and control and data acquisition equipment; (b) a secondary combustion chamber for allowing sufficient residence time at elevated temperatures to assure complete reactions and gasification of soot; this secondary combustion chamber can be fired either by a burner or by a low power non-transferred plasma torch; (c) depending on the waste, a water quencher, to avoid formation of dioxins and furans; (d) a cyclone or bag-house for particulate removal; (e) a scrubber for eliminating acidic gases; (f) a hydrogen sulfide absorber, if necessary; (g) high efficiency filters or precipitators for small particulate removal; (h) an activated carbon filter for removal of heavy metals; and (i) a fan for generating sub-atmospheric pressure in the entire installation. Additionally, waste preparation and feeding systems have to be integrated with the furnace. As the thermal plasma plant consists of a number of unit processes, a control system is essential to connect each process with others efficiently. In addition, a safety management system is necessary to protect workers (Heberlein and Murphy 2008).

The most important one is the gasification furnace equipped with a thermal plasma generator. As early as 1960s, Westinghouse produced plasma torches for the National Aeronautics and Space Administration (NASA) for its Apollo Space Program (Camacho 1988). In DC arc plasma, the plasma state is maintained between two electrodes of the plasma torch by electrical and mechanical stabilization that are built into the plasma torch hardware. Two arc attachment points are required to generate a plasma column: one attachment point at the solid-gas interface at the cathode electrode and another at the gas-solid interface at the anode electrode (Park et al. 2004) which are separated by an insulator. The high temperatures that occur at the attachment points of the plasma invariably promotes some vaporization of electrode materials at the attachment points. Hence, water cooling is used to minimize the rate of vaporization so as to increase the electrodes' lifetime.

Arc plasma torches can be classified as rod type and well type cathodes based on electrode geometry (Hur and Hong 2002). Thermal plasma torches can also be labelled as transferred and non-transferred types depending on whether or not arc attaches onto a substrate directly. Thermal plasma characteristics such as input power level, plasma flame volume, temperature field, velocity distribution, and chemical composition can be tailored to suit for each application.

4.4.2 Alter NRG/Westinghouse plasma gasification process

The Plasma Gasification Vitrification Reactor (PGVR) is a plasma-assisted gasification process commercialized by Alter NRG, who has recently acquired Westinghouse Plasma Corporation, the original developer and a leading supplier of non-transferred arc (NTA) plasma torch technology. The process is compatible with a variety of feedstocks such as MSW, RDF, ASR, petcoke with hazardous wastes and so on.

The wastes are injected into the top of Plasma gasification reactor which is refractory-lined internally to withstand high temperatures and the corrosive operating conditions within the reactor. The plasma torches are installed in the bottom of the furnace to enhance the melting of inorganic materials contained in solid wastes. Prior to it, a coke bed is formed inside the reactor which absorbs and retains the thermal

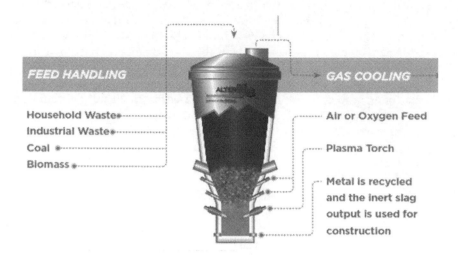

Figure 4.7a Alter NRG/Westinghouse plasma gasification reactor (Source: Westinghouse plasma corporation Brochure; reproduced with the permission of Westinghouse Plasma Corporation, Calgary, Alberta, Canada – mail from Business Analyst).

energy from the plasma torches for melting inorganic materials. The plasma torches are used to provide thermal power into the reactor, supplemented by heat released by the coke bed which is slowly consumed. The size of the standard Alter NRG/Westinghouse plasma gasification furnace varies from 4 to 9 meters top diameter and 10 to 24 meters overall height (Fig. 4.7a) (www.westinghouse-plasma.com/wpc_gasifiers_models).

The temperature of the plasma plume varies between 5000 and 7000°C and the bulk temperature within the base of the reactor is 2000°C (Helsen and Bosmans 2010).

Recently, a new design of furnace with a plenum zone (residence time ~2 sec) has been reported (Solenagroup website), which is a secondary combustion chamber for allowing sufficient residence time at elevated temperatures to assure complete reactions and gasification of soot. As in the case of AlterNRG reactor, in Solenagroup's furnace also, coke is added with the solid wastes, which is consumed in the furnace at a much lower rate than the waste material due to its low reactivity, and forms a bed onto which the MSW falls and quickly gasifies. The coke bed also provides voids for molten flux, slag, and metal to flow downward as the gas flows upward. The coke also reacts with the incoming O_2/steam to provide heat for the gasification of the feed materials. Oxygen/steam gasification improves hydrogen yield and delivers a high calorific syngas.

An LPG burner is installed sometimes in the furnace/reactor to add heat by raising the initial temperature when the heat value of solid waste is not enough. To lower the temperature of the gas (generally at >1200°C) exiting the furnace, a heat exchanger is installed behind the thermal plasma furnace to recover the heat from the gas. The recovered heat can be utilized to run a steam turbine. The cooled gases are then purified by *gas cleaning and conditioning processes* to generate clean syngas. The purified syngas can be used either to produce power (using a gas turbine or gas engine in an IGCC process) or high value chemicals or liquid fuels or high purity hydrogen.

Currently, the combination of thermal plasma process with methanation, or Fischer-Tropsch, or fuel cell with high purity H_2 are not in practice (Byun et al. 2013).

The molten slag is a mixture of non-combustible inorganics and recoverable metals which are sent to the slag handling system for further processing. Long residence times ensure cracking of any tars and thus minimize particulate matter. One advantage in using thermal plasma gasification is that NOx removal does not exist.

Helsen and Bosmans (2010) discuss the mass and energy balances for the Alter NRG core design with IGCC) for a typical American MSW (710 t/day) and tyre (40 t/day) feed. A 750 t/day plant would require about 6300 TPA of metallurgical coke which imposes considerable operating costs. Alter NRG also tested anthracite as a substitute for coke.

Most plasma arc facilities in Japan and North America are used for disposal of special industrial waste or hazardous waste. Some of these facilities provide thermal energy for district heating or generating small amounts of electricity. Due to the high temperatures generated by the plasma arc torches, these plants are used to dispose of such waste as asbestos, munitions, catalytic converters, aluminum dross, and fly ash. The capacity of these systems range from 1 TPD to 200 TPD, with most of them in the range of 10–20 TPD (Bowyer and Fernholz 2009).

Attempts to use plasma arc gasification to treat municipal solid waste have not been very successful, as it has a high moisture content as well as uneven composition. However, several semi-pilot test projects were done to evaluate the feasibility of plasma gasification of high calorific waste streams (such as RDF) and its impact on the environment (e.g., Lemmens et al. 2007; Circeo 2008).

4.4.3   Example of a thermal plasma facility

A Plasma arc gasification facility of 90 TPD capacity to treat MSW, set up in Ontario, Canada by Plasco Energy group is shown in Figure 4.7b.

This plant uses a more or less conventional reactor for the initial gasification of the solid waste. The final gasification of the residual char and vitrification of the bottom ash is carried out by a plasma torch. Plasma torches are also used to clean the raw syngas as it exits the reactor chamber and enters the cyclone. In this design, the syngas is cooled and used to run a reciprocating engine powered electrical generator. The heat recovered from the exhaust of the reciprocating engine, combined with that recovered from the cooling of the syngas can be used to generate low quality steam for district heating or bottom cycle power generation. This particular system experienced many operational issues including the need to build a waste water treatment plant onsite to treat the condensate recovered from the cooling of the syngas. The overall performance of the facility since the start of operation can be judged by the fact that on average less than 10 TPD was processed in its first three years of commercial demonstration although rated at 90 TPD (Wilson and Wilson 2013).

Westinghouse demonstration plant

Alter NRG (Westinghouse Plasma Corp.) owns and operates a demonstration facility (Westinghouse Plasma Center) located near Madison, PA, USA, which has a capacity of 48 TPD. The demonstration reactor was built in 1984 and several gasification tests

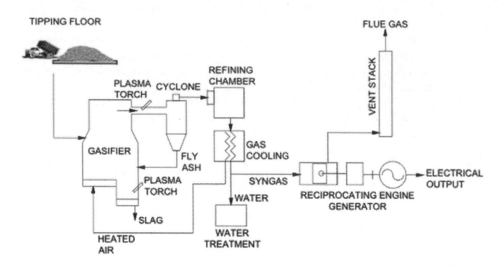

Figure 4.7b MSW gasification system employing plasma torches for slagging of the bottom ash and thermal cleaning of the syngas prior to entry into the cyclone and refining chamber (Source: Wilson and Wilson 2013; reproduced with the permission of Dr. Bary Wilson, Chief Oprations Officer, EPR).

were performed using over 100 different wastes including MSW, RDF, construction and demolition waste, hazardous waste, including PCB contaminated waste and harbour sediment sludge, waste water sludge, waste wood and clean wood chips, bagasse, tires, ASR, heavy oil, and incinerator ash.

Westinghouse continues to refine its core plasma torch and gasification technology based on the results obtained at its demonstration plant and the experience at operating facilities. The company announced that: At Tees Valley #1, Teesside, UK, a Plant to treat 'sorted' MSW using Westinghouse Plasma technology would be commissioned during 2015. The Plant first of its kind in UK was designed to have a capacity of 1000 TPD with Power – Combined Cycle to generate power (Projects/Westinghouse Plasma Corporation, at www.westinghouse-plasma.com/projects/).

4.4.4   Plasma technology for treatment of incinerator residues & hazardous waste

Small scale thermal plasma facilities applicable only to homogeneous waste streams and facilities treating MSW currently operating in Europe, the U.S. and Asia (especially Japan) are discussed in several publications (for example, Buyn et al. 2013).

The first generation of full-scale commercial PGVR systems to produce syngas from MSW, RDF and ASR has had a mixed success: one small facility (Mihama-Mikata in Japan, 25 tpd) is reporting nearly ten years of successful operation; a second large-scale facility (Utashinai 150–220 tpd) has been affected by a series of design flaws and operational issues, requiring re-engineering of major processing equipment and causing significant downtime. Ultimately the unit had to be shutdown in 2013 (http://www.westing house-plasma.com). Improvements are made from the lessons learned at Utashinai plant and the second generation of PGVR is being commercialized by Alter NRG Corp. Two of these new gasification facilities, 70 tpd, for hazardous waste treatment owned by SMS Envocare Ltd have been commissioned recently in Pune and Nagpur, India. SMS Envocare and Westinghouse Plasma Corporation have partnered to replicate the Pune plant configuration around the world, and over 30 projects are under different stages of development (AlterNRG website).

For treatment of *the incinerator residues*, plasma technology is extensively used since it offers distinct advantages (Heberlein and Murphy 2008). Besides reducing the volume of the waste typically to 50%, the toxic heavy metals can be encapsulated in a glassy slag formed with the addition of high temperature fluxing agents at typical slag temperatures, 1500 to 1700°C. This slag is non-leachable and the leach rates of heavy metals from these slags can be significantly lower than the specified emmissions limits.

Similar situation with contaminated soil and sewage sludge. These wastes can usually be treated at relatively low temperatures to remove contaminants and moisture, resulting in a volume reduction by a factor of about 20. Plasma can play a support role, in particular, in enabling the off-gases useful for combustion. Low level radioactive wastes (LLRW), asbestos and military wastes have higher negative values, and landfill disposal and incineration are in most cases not possible. In all such cases, plasma treatment yields a significant volume reduction, typically by a factor of 3. Besides, plasma processes may allow separation of hazardous materials (e.g. radioactive) from non-hazardous materials with different thermal characteristics. Control of the radioactive material in a non-leaching glassy slag has been achieved.

The low boiling point radioactive materials have to be carefully collected in a filter from the exhaust gases; also care must be shown in reprocessing. Several plasma processes have been developed for compaction of LLRW, and few of the processes are well established. Among the several approaches, plasma heating combined with either incineration or resistive heating should be mentioned. The first method is particularly suited when combustible exhaust gases generated are used for on-site generation of the electricity. In the second method, the waste is resistively heated by a current, thus using electrical energy more efficiently for the melting of the material, while the plasma provides the required high temperature environment.

The advantages of a combined plasma and incineration installation are described by Bendix and Hebecker (2003) who maintain that separating harmful components such as most plastic materials and metals containing halogens from the MSW would allow operation of an incineration plant at higher temperatures because less corrosion would be countered. In turn, this leads to higher temperature steam production and more efficient power generation. The harmful wastes would be treated in a parallel stream in a high temperature plasma reactor. A combination of conventional gasification technology with plasma treatment of residues is pursued by Tetronics of UK (Chapman et al. 1995, Deegan et al. 2002, Wolf et al. 1998, Chapman et al. 2007). For materials with a high fraction of metal content (e.g. some LLRW), combination of plasma treatment with induction heating may be applicable (Watanabe and Shimbara 2003).

Non-transferred arc processes are preferred for treatment of liquids and gases, because of the more uniform temperature distribution and the availability of reactive species. Further, the quantity of fluid requiring treatment is generally relatively low, so small installations are usually required, and non-transferred arcs are more easily

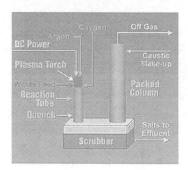

Figure 4.8 Schematic of the PLASCON process for the treatment of hazardous liquids and gases (Source: Heberlein and Murphy 2008; reproduced with the permission of Dr. A. B. Murphy, CSIRO, Australia).

scaled down. RF induction plasmas and microwave plasmas are also shown to be commercially viable in small-scale applications. To mention, Pyroplasma process of Westinghouse was used to completely destroy PCBs, and the PLASCON process was employed in Japan, Australia, U.S. and Mexico for destroying halons, CFCs, hydroflu-orocarbons, PCBs, insecticides and the waste liquid from herbicide manufacture (Heberlein and Murphy 2008).

The PLASCON process uses a dc plasma torch with a tungsten cathode with argon as plasma gas. The process has destroyed a wide range of hazardous waste liquids and gases, which are injected downstream of the plasma torch, together with an oxidizing gas such as oxygen or steam. After passing through the reaction tube, the hot gases are quenched and then scrubbed using caustic soda. A rapid water quench is used to avoid production of dioxins. A schematic of the process is shown in Fig. 4.8.

Typical feed rates are around 100 kg/h for halons and CFCs, for which destruction is to the level of 99.9999%. CHF3 (or HFC), which is a strong greenhouse gas with a very high global warming potential, is an unavoidable by-product of the process used to produce HCFCs. The Quimobasicos plant in Monterrey, Mexico, produces about 185 t/yr of CHF3, which is equivalent to 2.2 Mt/yr of CO2, corresponding to the emissions from a 300MW coal-fired power station or 500000 cars. The destruction of this toxic gas using the PLASCON process is comparatively a very low cost method. The PLASCON process has been well characterized by thermodynamic, chemical kinetic and fluid dynamic modelling studies, particularly its application to halons and CFCs, referred to collectively as ozone-depleting substances (ODSs). These studies are reviewed briefly by Heberlein and Murphy (2008).

Several other processes for destroying ozone-destroying substances are developed in Japan using plasma technologies (Watanabe and Shimbara 2003, Murphy 2003, Watanabe et al. 2005). These applications are highly appropriate in combating global warming.

4.4.5 Issues with plasma arc gasification

(a)   The stability and service life of the refractory linings in the reactor have been a problem in some designs because of the high temperatures involved. Temperature variations leading to thermal shock and attack of the liner material by highly reactive hot chlorine gas developed from solid waste can severely reduce refractory life.

(b)   As the electrodes which create the arc and generate the plasma are in constant contact with the solid waste, the characteristics of the waste as well as the corrosive gases created severely reduces the operating life of the electrodes. Thus, reactors are so oftenly removed from service to replace the expensive plasma electrodes or even to install torches. In some systems, the service life of these torches is of the order of 30 days between major component replacements. For this reason, pre-processing to obtain greater homogeneity and smaller size of the waste feed to the plasma reactor becomes a requirement to decrease the possibility of fouling or damage of the electrodes.

(c)   The high temperature gases produced from the feed stock can also include vaporized metals such as cadmium, mercury, and lead, and such contaminants as HCl, NH_3, H_2S, COS and so on which can be emitted into the atmosphere following the combustion chamber, internal combustion engines or gas turbine stage. By rapidly cooling the gas either before it enters the gas reactor or after, these metals can be captured and convert back into solid form. Cooling is difficult given the high temperatures generated by the plasma arc and, thus plants utilizing this technology face air pollution control issues. Quick cooling of the flue gases (exit gases) below 300°F within seconds is also necessary to prohibit the reformation of dioxins and furans that are destroyed during the incineration process. The plasma arc system does not have a boiler which can easily facilitate this rapid cooling, and the current designs employ a liquid scrubbing step. Currently, no data with respect to air quality for the plasma arc facilities in operation is available.

(d)   A tremendous amount of external, electrical energy is required to operate the arc which reduces the net positive quantity of electrical output from the plant. The different gasification approaches discussed are compared in Table 4.10.

The plasma arc gasification for MSW treatment has not been favoured mainly on economic grounds, despite superior technical merits (Bowyer and Fernholz 2009; Opinion Letter Regarding Plasma Arc gasification Options for management of Juneau's Waste SRS Engineers, 2009; www.gcsusa.com/pyrolysis.htm).

However, in practice, the plasma gasification process in combination with other technologies is utilized to recover energy from the syngas (e.g., Malkow 2004, Byun et al. 2011, Minutillo et al. 2009, and Galeno et al. 2011, Zhang et al. 2011).

4.5   COMMERCIAL STATUS OF GASIFICATION

Gasification has been used worldwide for almost two centuries to convert carbon-based materials such as coal, woodwaste, wood chips, and agricultural biomass to produce electricity and heat.

It can be a viable technology to process MSW as the individual processes involved were proven to work well (AES 2004). But, in the actual gasification systems of MSW, a large number of *uncertainties* (regarding performance, reliability and economics) are noticed. Much of the data on the performance of the few operational plants comes from

Table 4.10 Comparison of Gasification options.

Gasifier design	Tar in syngas	Cold gas efficiency	Operating energy requirement	Ability to handle variety of wastes	Permissible particle size	Moisture content (maximum)	Dust content
Downdraft	Low	>80%	Low	Moderate	<4 in	~40%	Medium
Updraft	Very high	>80%	Low	Low	<2 in	~50%	Low
Fluidized Bed	High	>90%	Moderate	Very low	<1/4 in	~10%	High
Plasma	Very low	>90%	High	Very high	NA	>50%	Low
Entrained flow	Very low	>80%	Low	Low	<1/25 in	~10%	High
Plasma enhanced Downdraft	Very low	>90%	Moderate	High	<4 in	>50%	Low

(Source: GBB 2013).

the individual companies which is inconsistent to compare between approaches. There are no papers that report on energy efficiency of commercially operating installations (except from the Energos installations). No public information is available on costs and long term performance. It is often unclear about the emissions involved, and the ash or other residue produced. Many of the gasification technologies have been mainly used for well-defined/specified waste streams, so their reliability and efficacy on mixed municipal waste is often debated. The ambiguity about the real practical and financial performance of these plants is resulting in preferring more established incineration technology. Equipment suppliers are not always financially sound and bid at too low prices, reducing project success chances in view of inferior interfaces (FOE 2009, ISWA 2013).

The gasification of MSW to generate steam or electricity has achieved different levels of commercialization in different parts of the world. In Europe, due to factors that include rising costs of landfills and higher taxes, the gasification option has become more attractive and several plants are operational. These are mostly fluidized bed type facilities built over the last nearly two decades (Jenkins, 2007). Facilities constructed in Germany and Italy were shut down because of economic and operational difficulties, and thus, the MSW gasification has been a mixed experience in Europe.

In Asia, particularly in Japan and South Korea, commercial scale gasification of MSW and industrial wastes has been made over the past 20 years (GBB 2013) and the performance has been good. Actually, Japan can be considered a leader in utilizing MSW gasification (see below). In the U.S. the technology was introduced in 1970s and '80s, when the MSW quality was more complex than today with low heating value, and hence not met with success. Recently, the technology has been reintroduced with improvements in the preprocessing of the municipal waste. Several commercial scale plasma arc plants have been proposed in the U.S. and some of them are under construction, although environmental concerns and a lack of trust in the technology are still cited (Wilson and Wilson 2013). But, a number of U.S. companies have pilot and demonstration facilities, and several commercial facilities are in advanced levels of development. Despite no commercial scale activity in the U.S., interest in gasification has grown in the last decade (GBB 2013).

From the point of installation and net operating costs as well as environmental impact, (air fed) gasification is still considered superior to pyrolysis, plasma arc gasification, and mass burn incineration (raw waste and RDF), whether or not combined with anaerobic digestion (Wilson and Wilson 2013). Also, the US Department of Energy (USDOE 2002) and the USEPA concluded that conventional, air fed gasification systems provided the most cost-effective and clean form of waste to energy systems, and that plasma arc systems were not cost- effective for municipal solid waste.

In summary, the operational experience of full-scale commercial gasification technology is very limited and divergent; thus, its overall efficacy is not well proven. Only in Japan, full-scale commercial gasification facilities processing MSW as a dedicated feedstock are currently operating. The uncertainies are still many to overcome before the technology is firmly established as alternative/or superior to mass burn Incineration. As regards thermal plasma gasification of MSW, although the technical feasibility of has been very well demonstrated, it is not clear that the process is economically viable globally, given the varying local conditions.

Commercial status of gasification in Japan

Based on the requirements, novel processes such as 'direct gasification and smelting', and the 'ash melting combined with rotary kiln or fluidized bed gasification' processes are developed and are operated in Japan.

There were about 30 Japanese companies engaged in the development of gasification and ash melting systems, which can be divided into three main types (Yoshikawa 2010): (1) The shaft types that melt the entire amount of wastes directly (e.g., JFE High Temperature Gasifying and Direct Melting Process); (2) The fluidized bed types that gasify the wastes directly with slagging (e.g., EBARA Fluidized Bed Gasification and Ash-melting Process); (3) The kiln types that gasify the wastes indirectly with slagging.

4.6 GASIFICATION PLANTS IN OPERATION

Most commercial processes use *combinations* of (pyrolysis/gasification/combustion) technologies. Nearly all of the commercial gasification systems actually follow the gasification – combustion or pyrolysis – combustion or all the three. In doing so many of the key theoretical differences between pyrolysis, gasification and incineration become blurred. While gasification is not the same as incineration, the actual practical differences between some 'commercial' gasification systems that incorprates incineration are relatively modest. Similarly, if pyrolysis is followed by combustion, there is relatively little difference from incineration. The principles of these novel processes along with their operational experiences are dealt by Malkow (2004) and Breault (2010).

The available conversion technologies have also been categorized as follows:

Low Temperature Conversion (LTC) technologies operating at <750°C that include slow pyrolysis and fixed-bed gasification technologies;

High Temperature Conversion (HTC) technologies operating at or above 750°C that include pyro-combustion, pyro-gasification and fluidized bed gasification technologies; and

High Temperature Conversion + Melting (HTCM) technologies integrating a very high temperature *melting* zone (above 1500°C) where minerals (ashes) and metals present in the waste stream are brought above their fusion temperature and recovered respectively as vitrified slag and molten granulates. These include plasma gasification, pyro-gasification + melting and fluidized bed gasification + melting technologies.

Many Companies worldwide are involved in the gasification, as technology developers, as facility developers and as both. A few of these have reached prototype/demonstration/commercial levels, and some are at advanced levels of development. Though the actual numbers and claims are not very clear, several publications provide the lists of the Companies along with the design and capacity of the gasifier, type of feedstock, emmissions data, years in operation and other details (e.g., GBB 2013, WSP 2013, Buyn et al. 2013, Foth Infra. & Environ. 2013, Themelis and Mussche 2013, RTI Intl. 2012, Williams 2012, E4Tech 2009, UC-Riverside 2012, PNNL 2008, Babtie Fichtner 2008, Klein and Themelis 2003). The Companies described differ from report to report, although a few are common. It is interesting to study these reports and company brochures, websites and presentations to realize why certain technologies are well estalished and others are not.

A few of the technologies are very briefly described here, selected on the basis of an advanced state of pre-commercial demonstration or commercial availability, and the number of plants currently operating. The material presented is drawn from the company brochures/websites or secondary sources.

(1) Mitsui R21 technology, Koga plant, Japan

The R21 technology for thermal treatment of MSW, developed by Siemens as Thermal Recycling Process was licensed in 1991 to MES. Currently Mitsui Babcock markets this technology.

The R21 technology combines rotary drum pyrolysis and high temperature combustion of the pyrolysis products (gas and coke) in a combustion chamber. The plant schematic is shown in Figure 4.9.

Technical data of the Koga plant: Number of lines: 2; Throughput: 2×130 TPD; Availability: 300 days/year; pyrolysis residence time: 1–2 hrs; pyrolysis temperature: 450°C; combustion temperature: 1300°C; Steam pressure and temperature: 4 MPa & 400°C; power generation capacity: 4.5 MW.

Process: The waste bunker is deep enough to store for five days capacity. A crane feeds the waste into a shredder where it is reduced to 20 mm size. A screw conveyor transports the shredded material into the internally heated 3.1 m wide and 23 m long rotary drum. The drum is sealed to prevent access of air in order to ensure an almost oxygen free atmosphere. The temperature inside the drum is 450°C, and the residence time of the waste is around 1–2 hours. During the process, the waste transforms into pyrolysis gas and a solid residue consisting of pyrolysis coke, inert matter, and metal scrap.

The typical concentration of the main species in the pyrolysis gas are:

O_{2-} 0 vol%; CO – 30 vol%; CO_2 – 20–40 vol%; H_2 – 15 vol%; CH_4 – 15 vol%; C_nH_m – 5 vol%; NH_3 – 20 to 100 ppm; HCN – 1 to 50 ppm; SO_2 <100 ppm; H_2S >1,000 ppm.

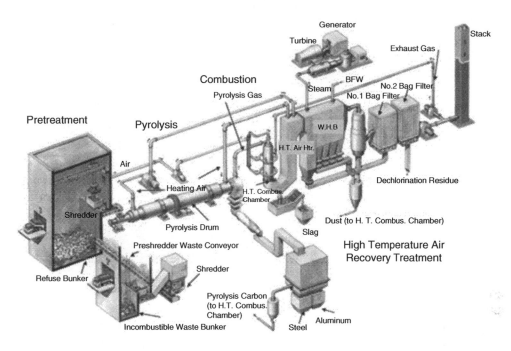

Figure 4.9 Schematic representation of R21 process (Source: Seifert and Vehlow 2009; reproduced with the permission of Dr. J. Vehlow, ITC).

The solid residues are discharged to a drum where they are cooled to approx. 80°C and are then transferred to a sorting stage equipped with a magnetic separator and an eddy current separation for the recovery of ferrous and non ferrous metals and inert materials. The residual fraction passes an air separator to blow the pyrolysis coke out. This coke is pneumatically transported into the high temperature combustion chamber where it is burnt together with the pyrolysis gas at a temperature of approx. 1,300°C. The inert residue can be utilised as aggregate in asphalt or concrete applications.

The liquid slag from the combustion chamber is quenched with water. The leaching of heavy metals out of this slag is very low, and the material can be utilised the same way as the inert residues from the pyrolysis process.

The hot flue gas leaving the combustion chamber enters the heat recovery system. In the first pass, air is heated which serves as a heat source for the pyrolysis drum, and the following part is a conventional boiler for generation of steam of 400°C and 4 MPa. This steam feeds a turbine to generate electric power, sold to a utility.

The air pollution control (APC) system starts with a fabric filter for removal of the fly ashes. Since this filter is considered as a dioxin removal one, the activated carbon is injected in the cylindrical container located between the boiler and the first filter. Next step is a dry process which uses freshly ground $NaHCO_3$ as a reagent, for reduction of acid gases.

The main advantage of the process is that it compensates the higher operation costs by the small amount of residues; the reaction products $NaCl$ and Na_2SO_4 account for about 95% of these residues, and the remaining 5% consist of other salts, including heavy metals. There is a good chance that heavy metal recovery and recycling of the alkali salts (offered in Europe by Solvay) will be economically attractive in the future (Seifert and Vehlow 2009: Appendix 2).

MES successfully built six plants in Japan operating on MSW with capacities of about 40000 to 120,000 TPA.

(2) Thermoselect process

The Thermoselect technology (licensed to Kawasaki Steel, now JFE Group of Japan and the Interstate Waste Technologies of the U.S.) is a gasification technology used to process MSW, commercial and industrial waste, medical waste, sludges, tires and electronic waste (Frank Campbell 2008).

The Thermoselect process (Miyoshi 1998) is a gasification plus melting technology which uses a gas reforming process to recover purified syngas from municipal waste and industrial waste. While minimizing environmental impacts, the process also realizes chemical recycling.

The process steps are the following: (1) The unshredded MSW is compacted to about one-fifth of its original volume using a hydraulic press; (2) drying and pyrolysis in externally heated tubes at about 600°C in the degassing channel, with waste feeding by ramming; (3) pyrolysis product is charged into a reactor where it is melted at high temperature (2000°C) with oxygen (gasification agent) and pyrolyzed carbon to form gas; (4) gas passing through the gas reforming/quenching/refining stages and recovering as a clean syngas.

The thermoselect systems require no waste preprocessing or RDF production. Water condensed from syngas cooling is treated for re-use as cooling water.

The merits of Thermoselect process are:

(1)   Very low emission of dioxins and no possibility of generation of fly ash: The generated gas is held at 1200°C for 2 sec or longer, and then quenched to about 70°C in an oxygen-free condition to suppress the generation of dioxins to an absolute minimum before recovered as fuel gas;

(2)   Complete recycling of wastes is possible: 100% of waste input is converted into pure syngas or recovered as granulated slag, metals, metal hydroxides, sulfur, mixed salts, and other substances which can be used effectively requiring no landfill disposal;

(3)   Clean gas can be recovered by gas reforming: Since the syngas is mainly composed of hydrogen and CO, the gas can be used as a fuel for power generation and/or as a chemical feedstock. The fuel gas is applicable to gas engine or fuel cell or gas-fired boiler or gas turbine combined cycle (CCGT) for power generation. This provides option to select an optimum generation method, given the equipment scale and site conditions;

(4)   The process is economically viable: The process utilizes the energy contained in waste to perform melting and eliminates the need for separate treatment processes for dioxins and fly ash with large fraction of heavy metals. As a result, the total cost is lower than the conventional incinerator that includes ash melting. In addition, as the landfill disposal is not required, the costs associated with constructing and maintaining landfills are saved (Yamada et al. 2004).

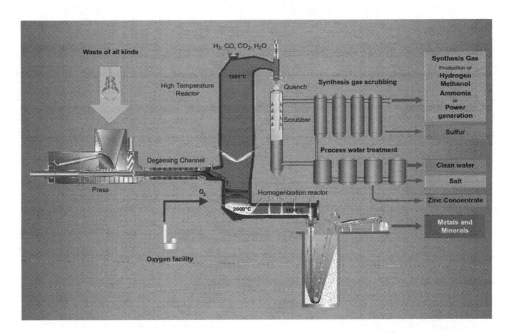

Figure 4.10 Schematic representation of a High Temperature Waste Gasifier based on MSW (Source: Frank Campbell 2008; reproduced with the permission of Francis Campbell, President, Interstate Waste Technologies).

Table 4.11 Data from Nagasaki Plant.

Emissions (mg/N-M^3 @ 7% O_2)	Measured	Japanese standard
PM	<4.7	15.4
HCl	11.6	126
NO$_x$	–	320
SO$_x$	–	225
Hg	–	–
Dioxin/Furan (ng/N-M^3)	0.025	0.14

(Source: UCR 2009).

Depending on the power generation cycle used, electrical efficiency of 11 to 40% may be achieved (Klein and Stahlberg 1995, Calaminus and Stahlberg 1998, Malkow 2004). Figure 4.10 provides a schematic representation of Thermoselect process.

The emissions results from the Nagasaki plant are shown in Table 4.11. Using gas engines and boiler, the Nagasaki Plant generates 8 MW electricity. The data given here is Compliance source test, June 2006.

Nine plants have been installed starting with a demonstration plant, 4.2 tons/hour capacity, in Italy in 1992. The Italy and Germany plants are closed now (Niessen, 2010). The seven Japanese plants are operating. This technology is not being offered currently by JFE due to unviable cost considerations.

JFE offers now Direct Melting Process (DMP) technology; based on this technology, ten plants are currently operating in Japan treating MSW, industrial waste, RDF and landfill minings. One plant is under construction in Albano, Italy to be opened in 2016 (ref: Proven Gasification Technology – JFE High Temperature Gasifying & Direct Melting Furnace; www.jfe-eng.co.jp). The Plant at Fukuyama near Hiroshima (world's largest gasifier) processes 314 TPD of pelletised RDF produced by seven units within the city area. The boiler produces steam at 60 bar and 450°C and exports 20 MWe from the steam turbine/generator. The flue gas is cleaned using a dry scrubbing system (lime + activated carbon) and a de-NOx process (SNCR). This plant is discussed in detail under Case studies in Chapter 5.

(3) Nippon Steel (direct melting) technology, Japan

The high temperature gasification process used by Nippon Steel, Japan is the most proven waste gasification technology with 35 years of operational experience. There are 42 plants operating in 35 countries (Ryo Makishi 2014).

The technology is based on a fixed bed updraft shaft-type gasifier. The process differs from conventional gasification by the addition of the ash melting stage. It is a high temperature gasification system, called Direct Melting System (DMS). The process produces a syngas that is combusted in a steam boiler, driving a steam turbine to produce electricity.

Technical data of the gasification plant located at Munakata city, Fukuoka prefecture:

Design waste throughput – 44,000 TPA; Number of lines – 2; Design waste throughput per line – 80 TPD; Design coke throughput – 61 kg/ton of waste; Waste throughput in 2008 – 32,177 tons; Design operation time – 240 d/a per line; Operation time furnace 1 in 2008 – 183 days; Operation time furnace 2 in 2008 – 202 days; Gasification residence time – 4 h; Gasification temperature – 300–1000°C; Oxygen in gasification air – 36 vol%; Combustion temperature ≈940°C; Steam pressure, temperature – 4 MPa, 400°C; Power generation capacity – 2.4 MW; Typical power generation – 1.2 MW (Seifert and Vehlow 2009).

Process: The MSW delivered by truck to the bunker is shredded for size reduction. The dried sewage sludge from a waste water treatment plant is mixed. A crane takes the waste mix to the top of the gasification furnace. The hazardous and medical waste is stored in a separate small silo from where it is also fed into the furnace.

For better control of the gasification process, a small amount of about 0.6% of coke is added to the waste in layers. For conditioning of the slag, as well as for capture of sulfur, lime stone is added. A schematic of the reactor showing the different reaction zones is given in Figure 4.11.

The waste enters the furnace via a double flap lock which ensures a tight separation of the furnace atmosphere from the surroundings. Oxygen enriched air (36 vol %) is blown into the furnace through four nozzles at the lower part. Air is injected through another six nozzles near the top. The energy for the gasification process is delivered by the combustion of residual carbon at 1,000–1,800°C in the lower part of the furnace. This high temperature causes melting of the slag which is drained through an outlet at the bottom of the furnace. The slag is then quenched in the water of the deslagger.

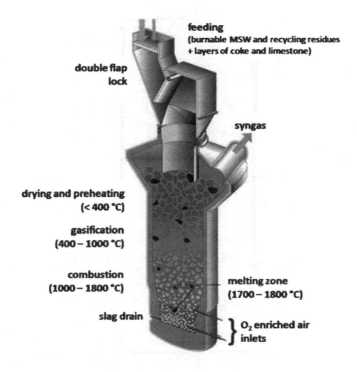

Figure 4.11 Schematic of the reactor (Source: Ryo Makishi 2014; reproduced with the permission of Mr. Ryoh Makishi, Nippon Steel).

After separation of ferrous and non-ferrous metals, the slag is used as construction material. The metal fractions are disposed for cost.

The gasification takes place in a temperature range between 400 and 1,000°C above the combustion zone. The hot syngas leaves the furnace after heating the mix waste, coke, and lime stone. The main components of the syngas are CO (15–20 vol%) and H_2 (2–5 vol%) and its calorific value is about 6.7 MJ/m^3.

In a cyclone directly behind the gas outlet, the coarse fly ashes are removed and returned to the furnace; the gas is then burnt in a cyclone combustion chamber at a temperature of around 940°C. The natural circulation boiler cools the flue gas down to 160°C. The boiler produces superheated steam at 39.2 bar and 400°C, which passes through a turbine generator to produce electricity. The turbine is a condensing unit with a rated output of 23.5MW. Nippon Steel claims an electrical efficiency of 23%. But the use of coke to generate the high temperatures for melting makes it misleading to compare this plant to a conventional incinerator or starved-air gasifier because a proportion of the input feed is a high calorific fossil fuel. Also, the parasitic load of the plant is relatively high considering the demand of the gasification process and O_2 enrichment plant. However, the main objective of thermal waste treatment in Japan is to minimize landfill and maximize resource recovery, and unlike many other regions such as Europe, energy recovery is a secondary consideration (WSP 2013).

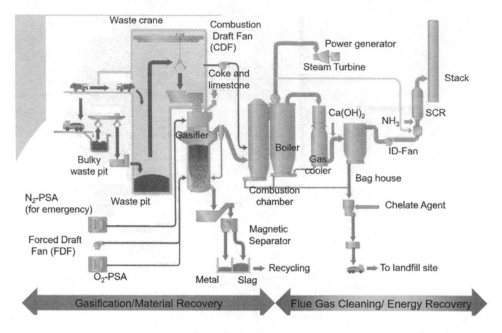

Figure 4.12 Schematic of Nippon Steel high temperature gasification technology (Source: Ryoh Makishi 2014; reproduced with the permission of Mr. Ryoh Makishi, Nippon Steel).

Most of the generated electricity is used in the gasification plant and in the adjacent recycling plaza, and only a small fraction is exported.

For air pollution control, dry scrubbing with injection of slaked lime into the flue gas is applied. Then the flue gas passes through a SCR system with ammonia injection into the reheated flue gas for NO_x removal before exiting through the stack.

The APC residues placed in bags are kept in a roofed concrete pit big enough to store quantities accumulated over five years. It is intended to recover heavy metals from this material in the future. The air emissions of the plant are in compliance with the emission standards (Seifert and Vehlow 2009). Figure 4.12 provides an overview of the high temperature waste gasification process (DMS) employed by Nippon Steel (Ryo Makishi, 2014).

The largest facility of this technology to treat MSW and sewage sludge is the Shin-Moji plant located in Kitakyushu city, Fukuoka Prefecture in Japan operating from 2007. The plant processes up to 216,000 TPA of MSW and sludge and has a capacity of 3 × 10 tons/hr. All the Nippon Steel DMS facilities world wide are listed with location, type of feedstock, number of lines, capacity per line and the annual capacity of the Plant in the WSP's report (2013).

(5) ENERGOS high temperature gasification technology

Energos ASA, Norway incorporated in 1995, designs and operates high temperature grate gasification technology that is cost-competitive, efficient, small-scale and environmentally friendly. The process combines gasification and combustion, and takes MSW or RDF.

The Energos patented software allows full control over the combustion process in a furnace of their design and creates a differentiated and more complete combustion reducing the need for expensive pollution cleaning systems.

Process: Drying, pyrolysis and gasification of the pre-treated waste is carried out in the primary chamber under sub-stoichiometric conditions. The syngas generated in the primary chamber is transferred to a separate secondary chamber where a final high-temperature oxidation takes place. The Energos furnace unit is horizontally divided into a primary chamber on the bottom, where the gasification of the solid waste takes place, and a secondary chamber on top of the primary chamber, where the combustion of primary gases is completed.

The received waste is pre-treated to ensure a sufficiently high surface-to-volume ratio, i.e., a particle size of <200 mm, with 90% of the shredded waste particles, <150 mm. The fuel bulk density after shredding and mixing should be >150 kg/m^3 and <500 kg/m^3. The content of other metals such as steel, stainless steel, iron and brass must be low, <0.5% in weight, and maximum particle size <40 mm after shredding.

In the primary chamber the waste is fed into the furnace in a controlled fashion, where it first falls onto a specially designed grate. At the input side of the primary chamber, drying of the waste takes place. Then follows a pyrolysis zone, followed by a carbon burnout zone at the hot end. The burnt out waste falls into a water bath/air lock and is removed and transported as bottom ash. The grate has no moving parts and its surface temperature is controlled. It is divided into twelve sections, and individually controlled air supplies provide primary air for each of the twelve grate sections. Over-fire air in the primary chamber provides an additional degree of freedom with respect to control of both combustion atmosphere and temperatures. The transport mechanism is designed in such a manner that in addition to the longitudinal transport there is good local mixing of the moving waste bed in order to promote the local homogeneity of the combustion process. After the combustible gases have left the primary chamber, secondary air and recycled flue gas is injected at several additional points, in order to achieve a proper combustion atmosphere and the right temperature trajectory. The furnace design makes it possible to simultaneously achieve good burnout of bottom ashes (and a low content of some heavy metals), low and stable CO and NO$_x$ emissions and a high degree of cracking of organic substances (Stein and Tobiasen 2004).

The Energos boiler system allows for rapid cooling of the flue gas. There are no cooled surfaces in the Energos furnace. The flue gas at a temperature, about 900°C enters the boiler. As the dioxins and furans are generally re-synthesized in the boiler system, a compact boiler system has been selected, based on a standard flue gas tube boiler design, followed by a standard water-tube economizer. In order to achieve rapid cooling and a compact design, the flue gas velocity needs to be markedly higher than the usual in traditional waste boiler systems.

The dry flue gas cleaning system is a standard baghouse filter with a high-performance membrane coating where lime and activated carbon are injected. Lime absorbs acid components (SO$_2$, HCL and F) in the flue gas while activated carbon absorbs TOC, heavy metals and dioxins. Dust/particles, lime and activated carbon will be separated from the flue-gas by the bag house filters. The fly ash in the gas stream is removed (see Figure 4.13).

The ENERGOS Process

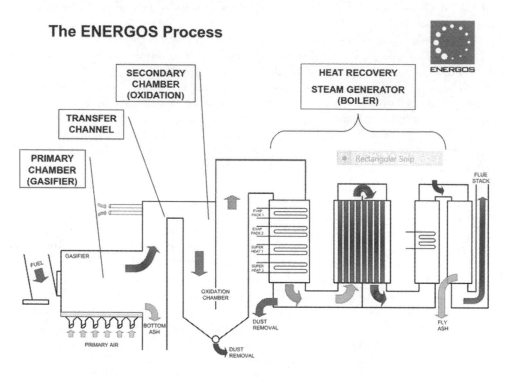

Figure 4.13 The ENERGOS Process (Source: Sandquist 2011; reproduced with permission from Judit Sandquist, Research Scientist, SINTEF).

The Energos Process Control system was designed to counteract variations in the waste feed, and thereby keep emissions below limits. The outer loop in the furnace control system controls the feed rate to the furnace by feedback control from the desired duty set point for the steam production in the boiler system. The inner loops control the addition of combustion air and recycled flue gas air at the various inlets. The control of the filter system (carbon and lime addition and filter pulsing pattern) is based on on-line measurement of the emission parameters to be controlled, with additional information relating the pressure drop across the filter system to basic filter characteristics.

Power and heat production

Energos offer energy recovery plants for CHP as well as power production by steam turbines.

Process residues: Water from boiler blowdown is used in the slag discharge basin. The reject from waste pre-treatment is sent to further re-cycling. Slag is typically used as topsoil at existing landfills and the filter dust is sent to special landfill sites (hazardous waste).

Stack emissions: Less than 0.1% of the emissions consist of harmful, polluting components. These emissions are well below the new EU emission requirements ranging from 1%–50% of the limits (Table 4.12).

Table 4.12 Emissions from the Plant.

Pollutant	ENERGOS (mg/Nm³)	EU limits (mg/Nm³)
Dust	0.24	10.00
Hg	0.00327	0.0300
Cd+Tl	0.00002	0.050
Metals	0.00256	0.500
CO	2	50
HF	0.020	1.000
HCl	3.6	10.0
TOC	0.2	10
NO$_x$	42	200
NH$_3$	0.3	10.0
SO$_2$	19.8	50
Dioxins	0.001 (ng TEQ/Nm³)	0.100 (ng TEQ/Nm³)

Source: Sandquist 2011.

Recovery of metals, including heavy metals: The metals entering a plant would to a large extent pass through the primary combustion chamber and end up in the bottom ash, partly oxidised. At the temperatures prevalent in the primary chamber, most of the metals have a negligible vapour pressure, so only a small fraction of them evaporate and follow the flue gas. Some of them, such as lead and zinc, may chemically react with substances with increased vapour pressure, and may be carried along with the flue gas. Minor entrainment of all metals as small metal particles may be expected. These metals are generally retained by the flue gas cleaning system. The mercury in the feed (being volatile) tends to vaporise and follow the flue gas. When the flue gas is cooled, more than 95% of the mercury and more than 99% of the cadmium (also volatile) condense or adsorb on dust and lime, and are retained in the flue gas cleaning system.

In a commissioned report to Energos on their plants operating within current operating limits, the components present in the feed that eventually end up as emissions to the air were estimated to be: Mercury: 2–5%, Cadmium: <0.01%, Arsenic: <0.03%, Cobalt: <0.05%, Nickel: <0.03%, and All other metals: <0.01%. The distribution of these components between bottom ash and filter ash may be manipulated to some extent by changing the temperature of the primary combustion chamber. Higher temperatures lead to less of the components in the bottom and more in the filter (Stein and Tobiasen 2004). The reported availability of the operating Energos plants is about 90%.

Commercial status: Six plants in Norway, one in Germany, and one in UK are successfully operating with different capacities ranging from 10,000 to 78,000 TPA. In addition, six new plants were reported to be in development in 2011 (Sandquist 2011).

(6) EBARA twin rec process, Japan

Ebara Corp. of Japan redesigned their well proven FB incineration technology to operate in a gasification mode which resulted in "Twin Internally revolving Fluidized bed

1	Fuel bunker	8	Bag house filter
2	Fuel crane	9	Filter residue silo
3	Hopper	10	Flue gas fan
4	Primary chamber (Gasification)	11	Chimney
5	Secondary chamber (High temperature oxidation)	12	Bottom ash extraction
6	Heat Recovery Steam generator (HRSG)	13	Steam turbine
7	Lime and carbon silo	14	Air cooled condenser

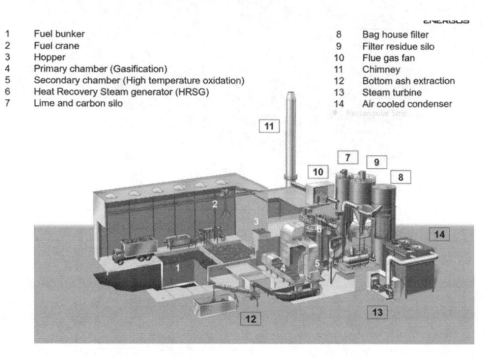

Figure 4.14 ENERGOS Energy from Waste Plant Layout (Source: Sandquist 2011; reproduced with permission from Judit Sandquist, Research Scientist, SINTEF).

Gasifier" process (TIFG). This process is combined with a technology for ash melting (Meltox) to produce a new concept for waste disposal (Figure 4.15). Based on this technology, a number of commercial plants were built, and a few are under construction.

An inclined distributor plate is used with a number of separate fluidising air supply chambers to provide differential air flows across the bed. In addition to promoting rapid and turbulent mixing of waste and bed material through the revolving action, heavy inert non-combustibles migrate to the sides of the bed for removal. A sloping furnace wall configuration just above the fluidised bed zone encourages the revolving action, restrains bed expansion and minimises bed carry-over. The controlled elliptical circulation patterns converge at the centre of the bed ensuring effective vertical and lateral mixing producing high combustion efficiency.

Shredded waste is fed into the hot revolving mass of bed material, typically silica sand, and sufficient air (sub-stoichiometric level) is injected from below to provide the fluidising action required to combust the organic materials. The organic compounds are transferred to the gas phase. The gasifier is operated at 500–600°C, low compared to the usual incineration temperatures.

Explanation of different components in the system shown in Fig. 4.16:

Ash Melter: Decomposes dioxins by high-temperature combustion (1,300°C~ 1,450°C);

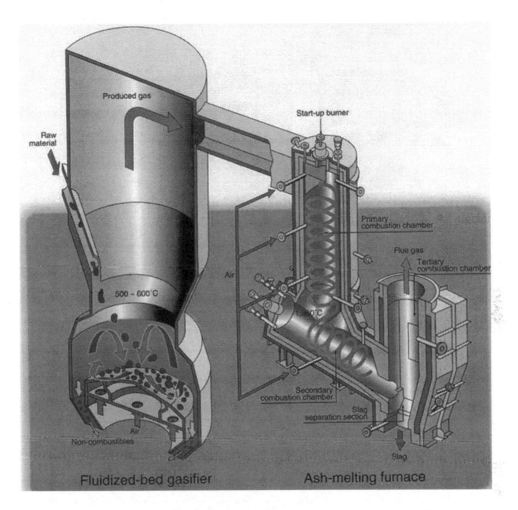

Figure 4.15 Schematic of the EBARA's revolving FB gasifier and the Cyclone combustion chamber (TIFG Process) (Source: Ebara Brochure; reproduced with the permission of Ebara Environmental Plant Co. Ltd., – Yasuhiko Hara).

Waste-heat Boiler: Recovers waste heat;

Gas Cooler: Cools inletgas to hazardous-gas remover to the temperature below 160°C

Bag Filter: Catchs molten fly ash;

Fly-ash Treatment Facility: Kneads cement and ferritization agent with water to stabilize fly ash, thereby preventing dissolution of heavy metals contained in molten fly ash;

Gas Scrubber: Alkaline solution is directly sprayed on exhaust gas to neutralize and remove HCl and SO_x. Heavy metals are also removed by liquid chelate agent. The scrubber condenses and removes moisture contained in exhaust gas;

Flue Gas Reheater: Heats scrubbed exhaust gas to an ideal temperature (180°C or higher) before going in to the catalytic reactor;

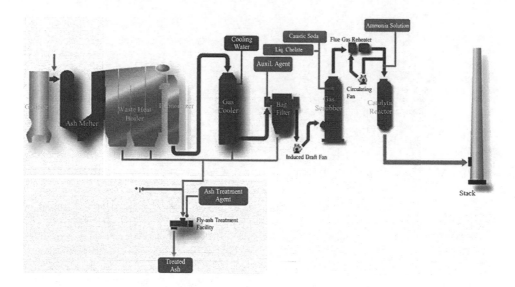

Figure 4.16 Flow diagram of EBARA process (modified and redrawn from Yoshikawa 2010).

Catalytic Reactor: Removes NO_x by catalytic reaction with ammonia as a reaction agent. The de-NO_x catalyst also efficiently decomposes dioxins.

Treated Exhaust Gas: The quantity of emission observed of each pollutant is the following. These are converted values assuming oxygen concentration as 12%.

Soot and Dust – 0.01 g/m^3 (NTP); Dioxins – 0.01 ng-TEQ/m^3 (NTP); HCl – 10 ppm; NO_x – 50 ppm; SO_x – 10 ppm.

The produced syngas and entrained particulates pass out of the freeboard zone of the fluidised bed and over into a cyclonic combustion chamber where they are subjected to combustion at about 1400°C by the addition of secondary air. The high temperature environment in the cyclonic furnace is sufficient to melt the inorganic ash components to produce a molten slag which is then quenched to produce a granulate material and removed. This material is claimed to meet the leachability regulations. Bottom ash from the gasifier is recovered from the base of the fluidized bed and the metal fractions are then separated by magnetic and eddy-current means into ferrous and non-ferrous fractions (iron, copper, and aluminium) for recycling. Since the fluidised bed is operated in gasification mode, the mass flow of the flue gas produced is small allowing the use of a compact sized steam boiler and air pollution control system. The energy content of the waste is converted via the steam cycle into electricity and/or heat energy.

The emissions data averaged across two months period in 2003 from the Toyohashi plant are given in the Table 4.13 (WSP 2013).

Commercially, 15 plants are in operation currently treating municipal solid waste and industrial waste, three of them in Korea (Yoshikawa 2010).

(7) InEnTech Plasma Enhanced Melting Process

The U.S. Company, **In**tegrated **En**vironmental **Tech**nologies, LLC developed the Plasma Enhanced Melter (PEM) combining plasma and glass melting technologies.

Table 4.13 Emissions to air from Toyohasi plant
(unit: mg/Nm³; * ng/Nm³ I-TEQ).

Pollutant	Regulatory limit	Measured values
Dust	<10	<1
SO₂	<28.6	<2.9
HCl	<16.3	<1.6
NOx	<98.2	29.7
CO	<12.5	2.5
Dioxins/furans*	<0.05	0.0073

(Source: Ebara).

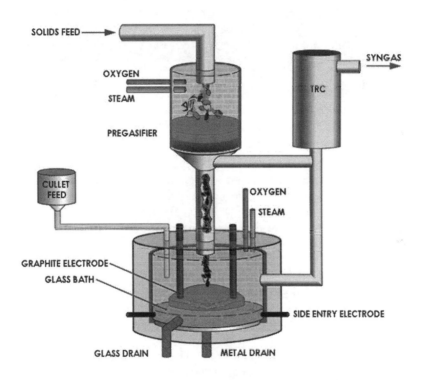

Figure 4.17 InEnTech's Plasma Enhanced Melter (Source: Jeff Surma, InEnTech Inc. 2012).

The system is designed to handle different types of waste: MSW, ASR and medical waste.

The overall system consists of downdraft pre-gasifier, plasma process vessel (PEM process reactor), and thermal residence chamber, the main unit being the downdraft gasifier (Fig. 4.17).

Here, the organic content of the feedstock is converted into gas. The remaining feedstock, including inorganic material and un-processed organics pass through the moving grate of the bottom of the pre-gasifier and into the PEM process reactor.

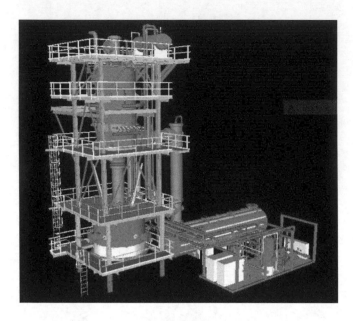

Figure 4.18   Commercial scale 125 TPD PEM design (Source: Jeff Surma 2012).

In the PEM reactor, there are two power systems: a DC plasma arc created between graphite elecrodes for high temperature organic waste destruction and gasification, and an AC powered joule heating system (JHS) to maintain an even temperature within the molten bath which facilitates the decrease of the power needed for the plasma arc. The plasma arc energy consumption is 74–86% of the overall energy input while the JHS is 14–21%. The plasma arc, thus, provides the heat to the process chamber, and the joule heating system allows even distribution of this heat within the molten bath.

In the PEM process chamber, the steam reacts with the volatilized organic portion of the waste to produce syngas, and the inorganics are vitrified to form a slag, which exits at the bottom of the molten bath. The syngas from the pre-gasifier as well as from the PEM process chamber enters the Thermal residence chamber (TRC) for about 2 sec when the hydroarbons are removed from the syngas.

The syngas leaving the TRC is cleaned and conditioned in a series of standard processes depending on the usage, either to produce power or to produce other secondary products (Ducharme, C 2010). There are 11 plants currently operating employing this technology (GBB 2013). The author's enquiry for the current status of these plants and technology has no response.

(8) The EDDITh process for small scale gasification

The process was developed by the French Institute for Petroleum (FIP), and currently commercialized by Thide Environment S.A of France. The Japanese Hitachi who was given the licence in 1999 have built several plants based on the technology.

The technology was specially developed for small-scale processing of MSW, industrial waste, RDF, ASR, electronic waste and sewage sludge. The Plant size ranges between 10,000 and 80,000 TPA.

The process consists of the following steps: The fuel preparation including grinding, screening and metals removal; Drying of the waste to 10–15% moisture level using a rotary hot air dryer, with hot gases supplied from the gasifier; Gasification of the waste in a rotating tubular pyrolysis reactor (kiln), heated to 450–700°C using the combusted syngas from the gasification process. The pyrolysis products are then discharged and the gas is combusted at around 1100°C in the combustor with air coming from the dryer. Then follows energy recovery from the hot flue gases for hot water, steam or electricity production.

The flue gas treatment includes activated carbon injection plus a bag filter. Scrubbing, sorting and processing of the carbonaceous solids from the gasifier are done to separate a solid residue (Carbor®), a coke/coal substitute with a heating value of about 16 Mg/kg, ferrous and non-ferrous metals, glass and inerts, and chloride salts (Fichtner 2008).

The Arthelyse plant in Frane consumes 40,000 TPA of domestic waste, 8000 TPA of general industrial waste and 2000 TPA of waste treatment sludge. The fuel moisture content was 31–44%, and had a LHV of 7.5–9.4 MJ/kg. The heating rate was 10–50°/min up to a final temperature of 400–700°C, and the residence time, 45–60 minutes.

Dry gas composition (data from Arthelyse Plant):

H_2 – 12.7 vol%; CH_4 – 16.0 vol%; CO – 19.1 vol%; CO_2 – 28.8 vol%; C_2H_4 – 5.5 vol%; C_2H_6 – 4.9 vol%; C3+ – 13.0 vol%; LHV – 23.1 MJ/kg (Stein and Tobiasen 2004).

Mass balance based on Arthelyse Plant:

1000 kg waste comprising 80% MSW, 16% industrial waste, and 4% sludge, with 220 kg of water out of dryer, 780 kg of dried waste to thermolysis process, yields 240 kg of solid fuel product (Carbor), 380 kg of thermolysis gas, 60 kg metals, 90 kg inerts,10 kg salt mainly $CaCl_2$ and NaCl. The Carbor solid fuel, the main product of the Process, represents approx. 45% of the waste energy content (Stein and Tobiasen 2004, Malkow 2004).

The process has been fully demonstrated at an industrial scale based on 500 kg/hr pilot plant in Vernouillet, France. There are three operating reference plants: (a) One plant in France operated by Thide Environment having a capacity of 50,000 TPA, processing household and small business waste, with steam exported to a local plant; (b) Two plants in Japan at Izumo and Itoigawa built by Hitachi: (i) Izumo: Capacity 70,000 TPA of household waste; outputs are electricity and Carbor® vitrified on site; (ii) Itoigawa: Capacity 25,000 TPA of household waste; Hot water output, with Carbor® used for combustion in a local cement plant.

The process schematic is shown in Figure 4.19 (Stein and Tobiasen 2004).

(9) Compact Power, Avonmouth, U.K.

The process conists of pyrolysis, gasification and complete combustion. The design temperature of the final flue gas stream is 1250°C in order to raise steam for CHP purposes.

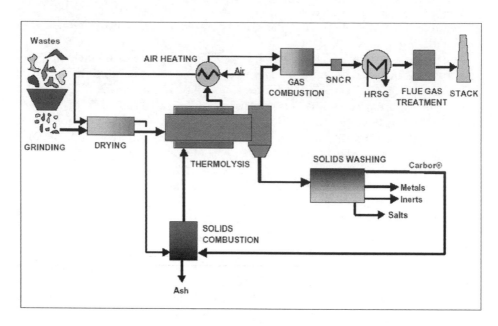

Figure 4.19 Schematic of the EDDITh process (Source: Stein and Tobiasen 2004; reproduced with the permission of Dr. Wes Stein, CSIRO Australia).

A special feature of the Compact Power design is modularity. A Compact Power facility would comprise multiples of a standard plant module, denoted MT2 designed to process 8,000 TPA of MSW. The advantage is not just ease of scalability (many gasifiers exhibit limitations to scaling up), but the front end of each module can be optimised to cater for a particular waste stream when the waste resource is mixed. This would involve adapting the feed handling system and controlling the pyrolysis chamber temperatures and residence times to suit each stream.

The plants have a nominal throughput of 6,000 to 30,000 TPA of waste, and energy can be recovered to generate heat and electricity.

Ethos Recycling took over Compact Power in 2008 and started the construction of a larger Avonmouth facility to process residual municipal waste. During 2014, CliniPower took over the ownership (Robert Eden, Waste Management World).

Process and Plant details: Bulky materials and recyclables are sorted out from the incoming MSW using a materials recovery facility. The waste is then placed into a hopper, via a tippler system, feeding a compactor/shredder. Compacted waste ensures that air is not drawn into the pyrolyser tubes through the feed port and the plant is kept under negative pressure so that all gases are drawn through the plant and processed under appropriate conditions. From the compactor, the feed is passed to a tube and screw pyrolyser. The pyrolysis chamber comprises two tubes, each with a screw feed. Each tube is approximately 3.54 m long with a diameter of 0.5 m, and can handle 500 kg/hr. Speed is controlled to give the material a residence time of approximately ½ hr; however screw speed can also be adjusted such that material

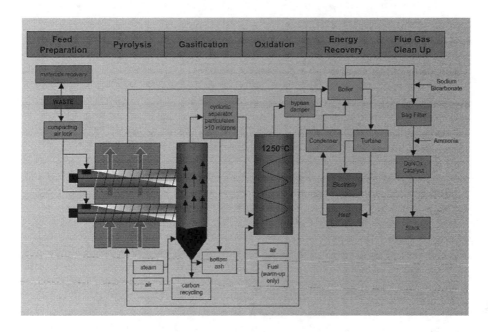

Figure 4.20 Schematic of MT2 module Gasification process (Source: http://www.surreycc.gov.
uk/__data/assets/pdf_file/0010/167437/Richard-Hogg-Compact-Power.pdf).

with variable calorific value can be accommodated (over a small range). The constant rotation of the material helps ensure good heat transfer through the feedstock.
The outside of the tube is engaged at a temperature of over 800°C. The waste passing through the pyrolyser is effectively carbonised, with volatiles being driven off,
leaving a carbon char. From the pyrolyser tube, the carbon char enters a gasification chamber, where it is reduced to ash by means of the water-gas reaction using
controlled flows of steam and air. The resulting syngas from gasification is combined
with the pyrolysis gas and passed directly into a thermal oxidiser, where it is combusted. As the gases are retained at high temperature, the condensation of tars is
not a problem as it happens generally in gasifiers feeding gas engines and turbines.
The thermal oxidiser is retained at a temperature of up to 1,250°C, with a retention time of two seconds to meet the usual technical standards. The exhaust from
the thermal oxidiser is then passed through a boiler to raise steam for CHP purposes.
Solid residues are removed as bottom ash that contain metals which are recovered
(Fig. 4.20).

Flue gas cleaning: A bag filter with sodium bicarbonate injection and selective
catalytic reduction (SCR) with ammonia for NO_x reduction are provided for cleaning
flue gas before exiting through the stack at 200°C. The emissions performance is
exceptionally good becauseof the good control of the gas combustion process and
the high temperature. The gas has a very fast residence time in the boiler (<0.2 s) to
minimize dioxin formation (Stein and Tobiasen 2004, Robert Eden). The estimated net

electrical efficiency of the process is about 14%. The process creates similar amounts of residues (or perhaps slightly more) as a conventional incineration process.

The MT2 module uses a 2-pass firetube boiler. In a MT8 plant, for an initial MSW stream of 4000 kg/hr having a calorific value of 12 MJ/kg, the flue gas available to the boiler is 29,440 kg/hr (mainly nitrogen) at 900°C. This could be used to raise about 11,500 kg/hr of steam at 350°C, 35 bar, and would generate about 2.2 MWe. The condenser would generally be air-cooled, partly to avoid the public misreading that the plume emanating from a wet-cooling towers is smoke.

Compact Power has a single reference plant operating under commercial conditions at Avonmouth, UK, with a capacity of 6,000 TPA. The plant was tested on RDF, unsorted MSW, sewage sludge, tires, food, paper sludge, leather although the facility currently processing clinical and pharmaceutical waste. Currently, the Plant appears to be not active.

Small-scale gasification & power generation – commercial demonstration

Relatively small and economically viable gasification technology to treat MSW that can be installed at distributed locations was developed by Yoshikawa and his group at Tokyo Institute of Technology. Three sizes of gasification plants with capacities ranging from 0.5 t/d to a few ten t/d were developed: small-scale (Micro STAR-MEET, 0.5–1 ton/day), medium (Mini STAR-MEET, several tons/day), and large-scale (Slagging STAR-MEET, several ten tons/day) to suit to onsite treatment of solid wastes generated by smaller towns (Yoshikawa 2004, Min and Yoshikawa 2004, Wang et al. 2004, Hara et al. 2004, Yoshikawa 2003, WSP 2013). This technology, if commercially exploited, may be quite pertinent to operate in towns with modest quantities of MSW generation without investing huge budgets.

4.7   ENERGY RECOVERY FROM PLASTICS

Worldwide 299 million tons of plastics were produced in 2013, a 3.9% rise over 2012. But recovery and recycling remain inadequate and millions of tons end up in landfills and oceans every year. Around 22–43% of the used plastic is landfilled globally. Most of the recovered plastics from the waste in the U.S. and Europe is received by China (nearly 56%) where it is reprocessed (mostly mechanical) in low-tech facilities with no environmental protection controls. Nearly 10 to 20 million tons of waste plastic end up in the oceans and this moving debris damage the marine systems and the local ecosystems (World Watch Institute website). The problems with used plastics or plastic waste can be reduced by recovering their fraction from the waste stream for effective recycling and energy generation.

Pure plastics are organic polymers which contain components with high molecular mass. Most of the plastics have petrochemical bases and are synthetic. There are over 20 different groups of plastics with different grades, but the main high volume commercial plastics are: polyethylene terephthalate (PET), high-density polyethylene (HDPE), low-density polyethylene (LDPE), polyvinyl chloride (PVC), polypropylene (PP), polystyrene (PS). PE, PP and PS are called thermoplastics that can be repeatedly soften and melt if enough heat is applied and hardened on cooling, so that they can be made into new plastics products.

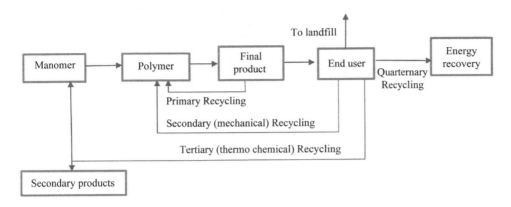

Figure 4.21 Municipal plastic waste Recycling methods (Redrawn from Achilias et al. 2012).

4.7.1 Recycling of plastic waste

Municipal Plastic Waste recycling technologies can be classified as primary, secondary, tertiary and quaternary recycling (energy recovery) as shown in Figure 4.21 (Al-Salem et al. 2009). Numerous publications have given excellent overviews of plastic waste management (Achilias et al. 2012, Butler et al. 2011, Castaldi et al. 2010, UNEP 2009, Panda et al. 2009, Al-Salem et al. 2009, Keane 2009, Psomopoulos et al. 2009, Aguada et al. 2008 & 2007, Themelis 2007 & 2013, Wong et al. 2015). Primary recycling involves processing of a waste into a product with characteristics similar to those of original product; secondary recycling (mechanical recycling) involves processing of waste plastics into materials that have characteristics different from those of original plastics product; tertiary recycling (thermo chemical) involves the production of chemicals and fuels from plastics waste as part of the municipal waste stream or as a segregated waste; quaternary recycling retrieves the energy content of plastic waste.

In this section, the tertiary and quaternary recycling aspects (thermochemical & energy recovery) which are fairly recent attempts are briefly discussed (Brems et al. 2013).

Waste plastics are the most available resource and also a most favourable means for fuel production because of its high heat of combustion. In Table 4.14 are shown heating values of various types of plastics along with those of other fuels, RDF, RPF, diesel oil, coal and so on. Compared to MSW and agriculture waste, plastics contain far lower moisture. Plastics are usually hydrocarbon-based and can be easily broken down into a liquid and further to a gas.

The conversion methods of waste plastics into fuel depend on the types of plastics utilized and the characteristics of other wastes/substances used in the process and the local conditions.

4.7.2 Thermal decomposition of plastics

Conventional combustion of plastics in waste incinerators may release their stored internal energy (Jinno et al. 2004, Zhuo 2009). However, direct combustion leads to diffusion flames around devolatilizing solids and to inefficient energy production.

Table 4.14 Heating values of combustible wastes
                and fuels (Kodera 2012 & Gao 2010).

Waste	HHV (MJ/kg)
Polyethylene	47.7
Polypropylene	45.8
Polystyrene	43.7
Poly (methyl methacrylate)	26.9
Poly (ethylene terephthalate)	24.1
RPF	>25
RDF	>12.5
Diesel oil	37.7
Heavy oil	39.1
Coal	25.7
Kerosine	43.4
Petrol	44.0
Liquid Petroleum Gas	46.1

Further, large amounts of health-hazardous soot, PAHs and other pollutants are generated (Goncalves et al. 2007). Burning PVC, particularly, create toxic polychlorinated dioxins.

Recently, much attention has been focussed to chemical recycling (mainly pyrolysis, gasification and catalytic degradation) as a method of producing various hydrocarbon fractions from plastic solid waste (PSW) because, by their nature, a number of polymers are suitable for such treatment. Thermolysis is the thermal treatment of plastic waste at controlled temperatures under a controlled environment; the processes can be either pyrolysis (thermal cracking in an inert atmosphere), or gasification (in the presence of sub-stoichiometric presence of air) and hydrogenation (hydrocracking) (Arhenfeldt 2007).

Pyrolysis can be successfully applied to PET, PS, polymethylmetacrylate (PMMA) and certain polyamides such as nylon, efficiently depolymerising them into constitutive monomers (Yoshika et al. 2004, Smolders and Baeyens 2004, Brems et al. 2007). Polyolefins, particularly PE, has been studied as a potential pyrolysis feedstock for the production of fuel (gasoline), or waxes as feedstock for synthetic lubricants, although with partial success. The development of value added recycling technologies is highly desirable as it would increase the economic incentive to recycle polymers. Several thermochemical recycling methods are currently in use: direct gasification, and degradation by liquefaction (Steiner et al. 2002). Various degradation methods for obtaining petrochemicals are presently under study, and conditions suitable for pyrolysis and gasification are extensively investigated (Aguado et al. 2007). Catalytic cracking and reforming facilitate the selective degradation of waste plastics. Solid catalysts such as silica-alumina, ZSM-5 or zeolites, effectively convert polyolefins into liquid fuel, giving lighter fractions as compared to thermal cracking. Gasification has recently been receiving more attention as thermo-chemical recycling technique. Its main advantage is the possibility of treating heterogenous contaminated polymers with limited pretreatment, and the production of syngas that has several applications in synthesis reactions or energy utilisation. Though gasification has been widely applied for biomass and coal, its application for the plastic solid waste treatment is less documented; nonetheless, the number of publications have been rapidly increasing (Brems et al. 2013).

Pyrolysis process is extensively used in the chemical industry for several applications, for example, to produce charcoal, to convert waste into syngas and bio-char and into valuable chemicals and for transforming medium-weight hydrocarbons from oil into lighter ones like gasoline. Pyrolysis of waste polyolefin plastics under different conditions can yield hydrocarbon waxes and oils, BTX aromatics, olefin gases (ethene, propene, and butadiene). The latter two groups of chemicals are particularly interesting since they comprise of the 6 base chemicals which are used as starting feedstocks in the synthesis of a huge array of chemicals consumed by the society (Chenier 2002).

Research on plastic thermal conversion

Significant amount of research on the subject has been done by several research groups (e.g., Welendziewski 2002; Manos 2006, Scheirs 2006, Kaminsky et al. 2004, 1995 & 1996, Kaminsky and Zorriqueta 2007, Stelmachowski 2010, Panda et al. 2009, Aguado et al. 2000, 2002, 2006 & 2008, Elordi et al. 2007 & 2009, Ates et al. 2013, Brems et al. 2013, Miskolczi et al. 2009 & 2013, Adrados et al. 2012, He et al. 2010, Butler et al. 2011, Bhaskar et al. 2003, 2006 & 2007, Williams 2006, Williams & Slaney 2007, Williams & Williams 1997 a & b, 1998, 1999, Achilias et al. 2006, 2007 & 2012, Gao 2010, Yoshikawa 2014). In these studies, the lab-scale pyrolysis reactors developed/employed were either batch type or semi-batch type. Most studies focused on the effects of operating temperature, heating rate, and catalysts on the final product yield. The pyrolysis reactor type significantly influences the heat transfer rate, mixing of plastics with pyrolysis products, residence time and the reflux level of the primary products.

The properties of pyrolytic products derived from different types of plastics vary significantly. PS and poly methymethacrylate have high monomer yields that approach 100% from pyrolysis, whereas PP yields only about 2% of its weight as its monomer (Kaminsky 1992). PE and PP contribute alkane and alkene respectively in pyrolytic oil from MSW pyrolysis, whereas PET contributes significantly to aromatic compounds (Cit 2010, Pinto et al. 1999, Onwudili 2009). Achilias et al. (2007) found from the pyrolysis of PP in a fixed bed reactor that the recovered oil and gaseous fractions contain mainly an aliphatic composition consisting of a series of alkanes and alkenes of different carbon number. These have a great potential for recycling back into the industry as a feedstock for the production of new plastics or refined fuels. Several papers are published on the studies made to assess the effect of the plastic mixture on product yield and composition to obtain the desired liquid product (Adrados et al. 2012, Pinto et al. 1999, Nilgun et al. 2004, Wu et al. 2014). Yan et al. (2015) studied the kinetics of virgin and waste PP and LDPE by a modified Coats–Redfern method and also investigated their thermal cracking in a semibatch reactor under atmospheric pressure in the presence of nitrogen. Both virgin and waste plastics were decomposed at 420–460°C. Due to the short residence time, the higher gaseous and liquid yields were obtained for virgin PP and LDPE. The high yields of gasoline (C_6–C_{12}) and diesel (C_{13}–C_{22}) fraction in liquid products confirm that this technique is desirable to realize waste plastics recovery.

Direct catalytic cracking of different type of plastics has been investigated by many using a large variety of catalysts. Zhao et al. (1996) have studied the effects of different zeolites such as H-Y, Na-Y, L, H-mordenite and Na-mordenite (their acidity favours cracking reactions) on the catalytic degradation of PP by thermogravimetry under nitrogen presence, and found that the degradation temperature of PP strongly depended on the type and quantity of zeolite used. Lin and Yang (2008) studied the catalytic cracking of plastic waste using zeolites. Adrados et al. 2012 found less expensive catalyst such as red mud, primarily consisting of Al_2O_3 and Fe_2O_3, promotes cracking and aromatization reactions that yield more gases and lighter, aromatic and fluid liquids. PE samples were also pyrolyzed using different catalysts, such as ZnO, MgO, CaC_2, and SiO_2 (Shah et al. 2010). He et al. (2010) explored the MSW and calcined dolomite pyrolysis in a bench-scale downstream fixed-bed reactor in the temperature range of 750–900°C. The results showed that the presence of calcined dolomite significantly influenced the product yields and gas composition. In the presence of calcined dolomite, the molar concentration of hydrogen is 66.30%, substantially higher than 36.69% obtained from non-catalytic pyrolysis processes, whereas CO_2, CH_4, C_2H_4, C_2H_6 concentrations decreased. The problem with the pyrolysis of PVC is the production of HCl and chloro-organic compounds. Two methods for reducing the influence of chlorine on the processes and utilization of products have been investigated (Miranda et al. 2001, Tiikma et al. 2006, Horikawa et al. 1999). In the first method, thermal degradation of plastic that contains PVC was done in two steps: the sample was heated to 320°C for dehydrochlorination and subsequently heated to a higher temperature for additional degradation. In the second method, HCl is fixed by adding absorbent, which is intimately mixed with PVC. These absorbents primarily include MaO, BaO, CaO, TiO_2 and Fe_2O_3. The Al-Zn composite catalyst (AZCC) was also found to act as a dechlorination sorbent as well as a catalyst in both the liquid and vapor phases (Tang et al. 2003). Blazso and Jakab (1999) observed that metals (aluminum, iron and zinc) and metal oxides (aluminum, titanium, copper and iron) reduce HCl formation and hinder benzene formation. The analysis of pyrolysis liquid from the pyrolysis of 'real' waste and 'simulated' waste composed of PE, PP, and PS plastics revealed significant amounts of styrene, toluene, and ethyl-benzene (Adrados et al. 2012, Zhou 2014). Donaj et al. (2012) studied how to increase the yield of the gaseous olefins (monomers) as feedstock for polymerization process as well as the applicability of a commercial Ziegler-Natta (Z-N): $TiCl_4$ /$MgCl_2$ for cracking a mixture of polyolefins consisted of 46%wt. LDPE, 30%wt. HDPE and 24%wt. PP. Two sets of experiments have been carried out at 500 and 650°C via catalytic pyrolysis (1% of Z-N catalyst) and at 650 and 730°C via only-thermal pyrolysis. A strong influence of temperature and catalyst presence on the product distribution was observed. The ratios of gas/liquid/solid mass fractions via thermal pyrolysis were 36.9/48.4/15.7%wt. and 42.4/44.7/13.9%wt. at 650 and 730°C respectively while via catalytic pyrolysis, 6.5/89.0/4.5 %wt. and 54.3/41.9/3.8%wt. at 500 and 650°C respectively. At 650°C the monomer generation increased by 55% as the catalyst was added. The yields of olefins were compared with the naphtha steam cracking process and processes for feedstock generation. Lin et al. (2005) investigated the catalytic cracking of PP in a fluidized bed reactor using H-ZSM-5, H-USY, H-mordenite, silica–alumina and MCM-41, with nitrogen as fluidizing gas and operating isothermally. The yield of volatile hydrocarbons for zeolite catalysts was higher than that for non-zeolite catalysts. Despite good performance by the catalyst cracking process, they can hardly be employed as of now, due to their high cost.

The decomposition of PE (both HD and LD) with pyrolysis temperature and the resulting yield of products among other aspects such as heating rate were studied by many (Mastral et al. 2006, Sorum et al. 2001, Conesa et al. 1994, Scott et al. 1990, Kaminsky 1980, Westerhout et al. 1998). Westerhout et al. observed that at 800°C, the product contains more methane than ethylene and low levels of aromatics irrespective of whether the type was HDPE or LDPE. (Mastral et al. 2006) studied the influence of residence time and pyrolysis temperature on the product distribution and found that waxes and oils could be obtained upto 700°C, and the gas yield increased with rise in temperature. Significant generation of aromatics were noticed at 800°C which increased with increasing residence time and temperature; the gas mainly contained hydrogen, methane and acetylene upto 1000°C. Lee et al. (2013) investigated the oxidative gasification of mixed plastics in a pilot-scale system and found the producer gas, after cleaning, mainly consisted of H_2, CO, CO_2 and CH_4. High temperature pyrolysis studies of PE and PP by several groups observed a mix of light hydrocarbon gases that included methane, ethylene, propylene, butylene, and so on (Westerhout et al. 1998, Zhuo et al. 2010, Costa et al. 2007, Faravelli et al. 1999, Sawaguchi et al. 1981 & 1980). Zhuo et al. (2010), however, found hydrogen also, in addition to the above consituents. Assumpcao et al. (2011) considered co-pyrolysis of PP with Brazilian crude oil by varying the temperature (400°C to 500°C) and the amount of PP fed to the reactor. The co-pyrolysis of plastic waste in an inert atmosphere provided around 80% of pyrolytic oil in which half represents the diesel oil. The studies further showed that the increase in temperature has favored the increase of pyrolytic liquid generation and the reduction of the solid formed. On the other hand, a huge increase in the quantity of PP has decreased total yield (liquid product). In general, it was observed that with temperature increase, there was a small reduction in yield in the diesel distillation range.

Several more interesting investigations can be found in a series of papers in the following volumes:

(a) Scheirs, J & Kaminsky, W (Editors), 2006: Feedstock Recycling and Pyrolysis of Waste Plastics: Converting waste plastics into diesel oil and other fuels, Wiley, New York, NY.

(b) Muller-Hagedorn, M & Bockhorn, H (Editors), 2005: Selected papers presented at Third International Symposium on 'Feedstock Recycling of Plastics & Other innovative Plastic Recycling Techniques' held at Karlsruhe University, Germany, September 2005.

4.7.3 Technologies for energy recovery from plastic waste

The types of plastics and their composition condition the conversion technology and determine the pretreatment requirements such as the combustion temperature for the conversion and the energy consumption, the output fuel quality, the flue gas composition (e.g. formation of hazardous flue gases such as NOx and HCl), the fly ash and bottom ash composition, and the potential of chemical corrosion of the equipment. Table 4.15 classifies various plastics and the types of fuel they produce. It can be

Table 4.15 Polymer as feedstock for fuel production.

Type of plastic feedstock	Description	Examples
Polymers consisting of carbon and hydrogen	Typical feedstock for fuel producton due to high heat value and clean exhaust gas	PE, PP, PS. Thermoplastics which melt to form solid fuel mixed with other combustible wastes and decompose to produce liquid fuel
Polymers containing oxygen	Low heat value	PET, phenolic resin, polyvinyl alcohol, polyoxymethylene
Polymers containing nitrogen or sulfur	Feedstock is a source of hazardous components such as NO_x or SO_x. Flue gas cleaning is required	Polymide and Polyurethane (contain N_2); Polyphenylene sulfide (contain sulfur)
Polymers containing chlorine, bromine, fluorine	Feedstock produces hazardous and corrosive flue gas	PVC, PVDC, Br-containing flame retardents and fluorocarbon polymers

observed that thermoplastics consisting of carbon and hydrogen are the most important feedstock for solid or liquid fuel production (UNEP 2009).

Plastic-to-fuel (PTF), also called Plastic-to-liquids is a set of nascent plastics-specific waste-to-energy technologies. PTF uses thermal processing techniques such as pyrolysis or gasification to convert scrap plastic to fuel sources such as syngas, oils, or liquid fuels by thermal or thermal/catalytic de-polymerization at moderate temperatures (e.g., Achilias et al. 2012, Butler et al. 2011; 4R Sustainability Inc. 2011). An external heat source is needed and is usually provided by combustion of the product gas in a separate combustion chamber and then by transferring the heat of combustion to the pyrolysis reactor across a metal interface. The resulting products are then upgraded to higher quality fuels that are more compatible with current fuel handling infrastructure designed for natural gas, gasoline, or diesel fuels. Gasification typically requires more heat and drives the chemical reactions by controlling the oxygen quantity in the system to produce syngas that can then be converted to methane. Pyrolysis operates at lower temperatures and in the absence of oxygen, and the pyrolysis products require more steps to refine into high quality fuels (Gendebien et al. 2003).

The recovery of high-value materials from waste pyrolysis include the recovery of carbon fibres from plastic composites for use in the aerospace and automotive areas. Pyrolysis of the the composite plastic results in oil, gas, and residual carbon fibre and some char. The char is amorphous and can easily be separated from the carbon fibre by mild oxidation, producing a recovered carbon fibre with strength properties nearing 95% of the properties of the original material. Steam gasification has also been used to upgrade the char product from pyrolysis of a range of waste materials, resulting in the production of activated carbons. Wastes such as tires have also produced activated carbons with similar surface areas and porosities to those of commercial-grade activated carbons (Williams 2012).

A couple of laboratory-scale experiments, designed and performed as part of technology development are now briefly discussed.

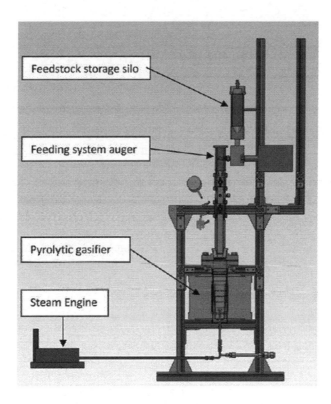

Figure 4.22 Lab-scale Experimental setup (Source: Soheilian et al. 2013; reprinted with the permission from Professor Y. Levendis and from ACS, © copyright 2013 American Chemical Society).

4.7.3.1 Laboratory-scale pyrolytic gasification technique

Levendis and his group at Northeastern University is one of the several investigators who studied extensively the pyrolysis of waste polyolefins. The objective of the Lavendis group was to produce a high- energy-content *gaseous fuel* from waste polyolefin feedstocks in an environmentally friendly manner (Soheilian et al. 2013, Caponero et al. 2005, Ergut et al. 2007, Goncalves et al. 2007a, 2007b). This gaseous fuel can be utilized for process heat, power generation, and other energy related applications. Polyolefins have the most popular commercial uses because of their excellent chemical resistance, versatility and low-cost. The entire system is shown in Figure 4.22. Its major components are a feedstock storing silo, feeding system, a heating chamber, and a steam engine.

This system was designed to pyrolytically gasify pelletized polymers at steady-state steady-flow conditions generating a combustible gaseous fuel, when burned in the miniature steam engine generates electricity. N_2 and CO_2 were used as carrier gases for the process to occur. This combustion system could burn bulk of the gaseous fuel generated by the pyrolyzer. A slipstream of the effluent of the very same combustion process was used to supply the pyrolyzer with a mixture of N_2 and CO_2 gases, upon removal of combustion-generated H_2O in a condenser. There are other additional benefits in using CO_2, as observed by Yamada et al. (2010), in the catalysis-aided reforming of the pyrolyzates of PE to CO and hydrogen at 950°C.

The furnace chamber is a SS tube with a heated volume of nearly 3 liters, suficient enough to accommodate the expansion of the pyrolyzing polymers fed with a mass feeding rate of 1 g/min, and the carrier gas flow at 1 L/min. A PID loop temperature controller (ATS Series XT16) was added for precise and reliable regulation of the system temperature. The speed of the electric motor that drives the feeding system was adjusted to the desired mass flow rate of plastic pellets. Asymmetrically perforated steel plates inserted at several vertical locations in the tubular furnace intercepted the falling polymer pellets and facilitated their melting and gasification in sequence. Thus, the particles gasified in the radiation cavity of the furnace without settling at its cooler bottom. The gas exit tube was fixed at the top of the chamber to prevent its plugging by tars, and also protected by a conical roof to prevent molten polymer ligaments falling on it. A gas temperature of 800°C was maintained to maximize the yield of gaseous PE pyrolyzates while minimizing the generation of tars and oils.

Commercially available pellets of post-consumer PP (density range, 0.86–0.95 g/cm^3) and LDPE (density range, 0.91–0.94 g/cm^3) were used as feedstock. The results showed that the major pyrolyzates in the gaseous fuel product were ethylene, methane and hydrogen. While ethylene content was maximum in the LDPE pyrolyzates, methane content was maximum in PP pyrolyzates. With a polymer feeding rate of 1 g/min, the flame obtained could generate steam and run the miniature steam engine at 1800 rpm sustaining the boiler pressure of 1 bar for about 20 minutes and produce electricity to power a small light bulb. The results of these studies and their analysis along with numerical approach are well discussed in the publication by Soheilian et al. (2013). These laboratory-scale studies clearly established that high-energy-content gaseous fuels can be produced from waste plastics which can be utilized for electricity generation.

4.7.3.2   Laboratory-scale pyrolysis of segregated plastic waste

The research at School of Engineering of Bilbao in Spain showed that pyrolysis of the segregated plastic waste generates *diesel fuel and syngas* (Adrados et al. 2012; Lopez et al. 2012). The process could be catalytic or noncatalytic with single step or multiple step process. The single stage uses a constant temperature for the reaction while the multi steps involves temperature rising in intervals during the pyrolysis. Earlier studies by López et al. (2011) on pyrolysis of plastic wastes showed that with increase in pyrolysis temperature, gas yields significantly increase while the liquid yields reduce. Also, the optimal temperature in terms of both conversion and the quality of the pyrolysis liquids was found to be 500°C when the char yield was insignificant.

The combination of catalytic and stepwise pyrolysis was applied to study the thermal conversion of a *mixture* of plastics (Lopez et al. 2012). A *mixture* of PE, PP, PS, PET, and PVC which resembles real municipal plastic waste was pyrolyzed in a 3.5 dm^3 semi-batch reactor at 440°C for 30 min using a ZSM-5 zeolite catalyst. A low temperature (300°C) dechlorination step was carried out, with and without catalyst, and was shown that the application of such dechlorination step gave rise to a 75 wt% reduction of chlorine in the liquid fraction. However, such step had a negative influence on the catalyst, loosing some catalytic activity. The studies found that, in terms of quality

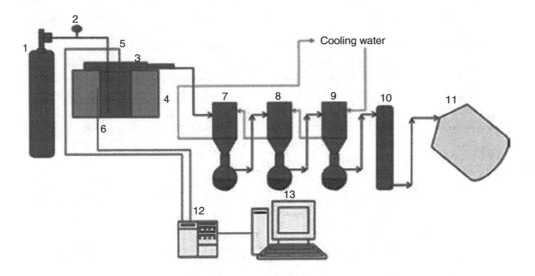

1. Nitrogen cylinder, 2. Rotameter, 3. Reactor, 4. Furnace, 5.Reactor thermocouple,
6. Furnace thermocouple, 7. Condenser 1, 8. Condenser 2, 9. Condenser 3,
10. Activated carbon column, 11.T bag, 12. Temperature controller, 13. PC

Figure 4.23 Schematic of the experimental system for Pyrolysis of plastic mixture (Redrawn from Lopez et al. 2012).

and chlorine content of the products, the desirable procedure was to pyrolyse the mixture at low temperature without catalyst and then performing the catalytic pyrolysis. A schematic of the experimental set-up is shown in Figure 4.23.

The rejected streams of packing and packaging plastic waste, separated and classified at materials recovery facilities, consist of many different materials, e.g., PE, PP, PS, PVC, PET, acrylonitrile butadiene styrene (ABS), aluminum, tetra-brik, and film, and to completely separate them is not technically possible or economically viable. Therefore, they are usually sent to landfills or incinerators. Adrados et al. 2013 have characterized and separated these materials into families of products of similar nature in order to determine the influence of different types of ingredients in the products obtained in the pyrolysis process. The pyrolysis experiments were carried out in a non-stirred batch $3.5\,dm^3$ reactor, swept with 1 litre/min nitrogen at 500°C for 30 min. Pyrolysis liquids were composed of an organic phase and an aqueous phase. The aqueous phase was greater as the cellulosic material content in the sample was higher. The organic phase contained valuable chemicals such as styrene, ethylbenzene and toluene, and had a HHV of 33 to 40 MJ/kg. Therefore they could be used as alternative fuels for heat and power generation and as a source of valuable chemicals. Pyrolysis gases were mainly composed of hydrocarbons but contained high amounts of CO and CO_2; their HHV was in the range, 18–46 MJ/kg. The amount of CO and CO_2 increased with consequent decrease in HHV as the cellulosic content of the waste was higher.

Pyrolysis solids were consisted of inorganics and char formed in the process. The presence of cellulosic content was found to lower the quality of the pyrolysis liquids and gases and increase the production of char.

Earlier, Adrados et al. (2012) reported results of a study on the pyrolysis of *simulated* plastic mixture and *real* waste sample from a sorting plant for comparison using the same non-stirred semi-batch reactor. Red mud, a byproduct of the bauxite refining, was used as a catalyst. Despite the fact that similar volume of these samples used, there were notable differences in the pyrolysis yields in terms of the compositions. The real waste sample resulted in higher gas and solid yields and less liquid fraction.

Other lab-scale experiments

Pinto et al. (2002, 2003) studied the fluidized bed co-gasification of PE, pine and coal and biomass mixed with PE. Slapak et al. (2000) designed a process for steam gasification of PVC in a bubbling fluidized bed. Xiao et al. (2009) co-gasified five typical kinds of organic components (wood, paper, kitchen waste, PE-plastic, and textile) and three representative types of simulated MSW in a fluidized-bed (400°C–800°C). It was determined that plastic should be gasified at temperatures in excess of 500°C to reach a lower heating value (LHV) of 10 Mg/Nm^3.

Brems et al. (2012) performed gasification of PET, PE, PP and PS waste using their chips (1–2 cm long and below 0.5 mm thick) or as pellets of about 3 to 5 mm in a bubbling fluidized bed reactor. The air flow rate was maintained at about 20% to 25% of the stoichiometric air flow needed for the combustion of each specific plastic material (Equivalence Factor, EF, 20%–25%), and the velocity in the bed varied from 2 to 4 times the minimum fluidisation velocity. For some of the experiments, a steam dilution was also applied, reducing EF to below 15%. About 0.3 to 0.5 kg of the feedstock sample was placed into a basket with perforations to enable the fluidized bed sand to fluidize within the voidage of the feedstock chips. The fluidized bed consisted of about 15 kg of quartz sand (150–300 μm, 2600 kg/m^3), for a total static bed height of about 35 cm. After appropriate analysis of CO, CO_2, H_2 and CH_4, the gases were exited to the atmosphere. Temperatures between 550°C and 800°C were tested. Experiments were also performed where live-steam (110°C, 5 bar) was directly injected into the fluidization air flow, further reducing the partial pressure of O_2 present. The results were used to determine the kinetics of the plastic waste gasification. The progress of the gasification reaction was monitored through the on-line measurement of the CO production. A first order gasification reaction was assumed to be valid, allowing the calculation of the reaction rate constants from plastic waste conversion as measured by CO production. The order of the reaction was determined by Brems et al. (2011), and indeed found to be close to 1 for all polymers. From the results, the reaction rate constant was determined using the Arrhenius equation for these plastic wastes.

Large-scale processes

One of the most common large-scale technologies is the (WGT) process developed by Waste Gas Technology Ltd., UK. The plastic feedstock is dried and mechanically pre-treated, sorting out incombustibles and granulated to optimum sized particles and fed into a cylindrical reactor for gasification at 700°C–900°C to yield a high calorific value gas (WGT 2002). Upon discharge and subsequent separation of gas and char, the latter may be utilized via combustion in a boiler to raise steam while the gas is quenched and cleaned of contaminants prior to its use in a gas engine or turbine and possibly CCGT applications.

The Texaco Gasification process is by far the most well known technology. First pilot scale experiments (10 tons/day) were carried out in the U.S (Weissman 1997). The process consists of two parts: a liquefaction step and an entrained bed gasifier. In the liquefaction step the plastic waste is mildly thermally cracked into synthetic heavy oil and some condensable and non-condensable gas fractions. The non-condensable gases are reused in the liquefaction as fuel (together with natural gas). Oil and condensed gas produced are injected into the entrained gasifier (Croezen and Sas 1997). The gasification is carried out with oxygen and steam at about 1200°C to 1500°C. After a number of cleaning processes, a clean and dry syngas is produced, consisting predominantly of CO and H_2, with smaller amounts of CH_4, CO_2, H_2O and some inert gases (Tukker et al. 1999).

To achieve break-through of these processes, further experimental work to improve the equipment design and product optimization is required. Advances in that area will facilitate in the improvement and more widespread use of gasification reactors.

4.7.3.3 Demonstration-level liquid fuels production from plastic pyrolysis

As the chemical energy stored in one ton of NRP is equivalent to about 5 barrels of oil, the processes, operating at 80% thermal efficiency, are expected to produce, on average, 4 barrels of oil per ton of plastic wastes processed (4RS Inc. 2011, URS Corp. 2005, Themelis & Mussche 2014).

The pyrolytic action of the plastics and the condensation of the resulting hydrocarbons results in the production of liquis fuels. The plastics suitable for the conversion (such as PE, PP and PS) are introduced into a reactor where they decompose in an inert gas ambient at 450 to 550°C. Depending on the pyrolysis conditions and the type of plastics, carbonous matter gradually develops as a deposit on the inner surface of the reactor. After pyrolysis, this deposit should be removed to maintain the heat conduction efficiency of the reactor. The resulting oil, a mixture of liquid hydrocarbons, is continuously distilled after the waste plastics inside the reactor are decomposed enough to evaporate on reaching the reaction temperature. The evaporated oil is then cracked with a catalyst (such as silica-alumina or zeolite). The boiling point of the produced oil is controlled by the operation conditions of the reactor, the cracker and the condenser. Sometimes, fractional distillation is needed to meet the user's requirements (UNEP 2009, Kodera 2012). After the resulting hydrocarbons are distilled from the reactor, some hydrocarbons with high boiling points such as diesel, kerosene and gasoline are condensed in a water-cooled condenser. The liquid hydrocarbons are then collected in a storage tank through a receiver tank. Gaseous hydrocarbons such as methane, ethane, propylene and butanes cannot be condensed and are therefore incinerated in a flare stack. This flare stack is required when the volume of the exhaust gas emitted from the reactor is large. The schematic diagram of a liquid fuel production plant using a tank reactor is shown in Figure 4.24.

The feeding methods vary depending on the characteristics of the waste plastic. The easiest way is to introduce the 'untreated' waste plastics into the reactor. Soft

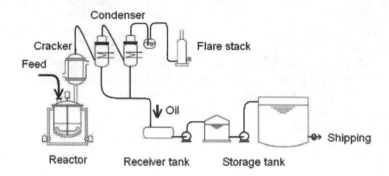

Figure 4.24 Typical layout of a liquid fuel production with a tank reactor which most recyclers in Japan who commercially produce liquid fuel from plastics of industrial wastes utilize along with a distillation system (Source: Kodera 2012, INTECH, Open access).

plastics such as films and bags are often treated with a shredder and a melter (hot melt extruder) to reduce their volumes before feeding them into the reactor.

Different types of reactors – tank, screw, externally heating-kiln, and fluid bed types – and heating equipment are used. Also, induction heating by electric power has been used as an alternative to using a burner.

The formation of carbonous matter in the reactor acts as a heat insulator; as such in some tank reactors a stirrer is used to remove the carbonous matter rather than for the purpose of stirring. After the liquid product of the pyrolysis is distilled, the carbonous matter is taken out either with a vacuum cleaner or in some cases reactors are equipped with a screw conveyor at the bottom of the tank reactor to remove the carbonous matter.

The amount and composition of the waste plastics are very much related to the operating conditions. The Plant performance is evaluated considering the energy consumption and plant costs relative to the plastic treatment capacity as the the typical criteria. Most important are the safety considerations in this type of chemical conversion because of the formation of highly flammable liquid fuels.

Yoshikawa and his group have been investigating to produce both liquid and gaseous fuels from waste plastics for power generation (http://yoshikawa-lab.org). They have proposed a sequential pyrolysis and catalytic reforming system for waste plastics degradation over commercial catalyst and modified natural zeolites. This novel system utilizes diesel fuel and gaseous product together for fueling a dual-fuel diesel engine to generate electricity. The solid products will be used for co-combustion with coal and biomass. The process flow diagram is shown in Fig. 4.25.

The experiments and resuts are discussed in detail by Syamsiro et al. (2013, 2014). The PP and PS samples were degraded at 500°C in the pyrolysis reactor and then reformed at 450°C in the catalytic reformer. The results showed that mordenite-type natural zeolites, either calcinations treatment (A-NZ) or HCl treatment (H-NZ) or nickel impregnation (Ni-NZ) could be used as efficient catalysts for the conversion of PP and PS into liquid and gaseous fuels. The high quality of gaseous product can be

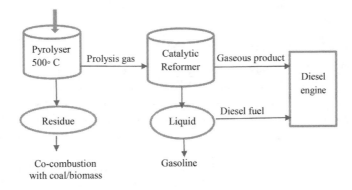

Figure 4.25 A schematic diagram of the proposed system for converting waste plastics into fuels (Redrawn from Yoshikawa Lab. Website).

Table 4.16 Properties of oil produced from polyethylene bag wastes.

Properties	Units	No catalyst	Y zeolite	Natural zeolite	Diesel fuel	ASTM method
Density	g/cm³	0.8719	0.824	0.868	0.8445	ASTM D1298
Kinematic viscosity	mm²/s	1.999	1.838	2.191	1.3–4.1	ASTM D 445
Pour point	°C	27	24	24	6	ASTM D 97
Heating value	MJ/kg	46.74	46.67	45.58	46.5	ASTM D 240

used as a fuel for driving gas engines or for dual-fuel diesel engine. The properties of pyrolysis oil produced from polyethylene bag wastes are given in Table 4.16.

Liquid fuel can be a substitute for liquid petroleum. The resulting products and fuel quality depend on the plastic types and decomposition conditions. Polyamides and polyurethane give oily products of high nitrogen content at low yields while PET gives solid products including terephthalic acid.

The solid product, carbonous matter can be used as a feedstock for solid fuel. Aluminum foil or other inorganics may also be present depending on the waste composition.

Pilot scale experiments utilizing municipal plastic wastes as feedstocks have been done in cooperation with Gadjah Mada University, Indonesia. The performance of the engine has been tested using the produced plastic oils as fuels.

The performance test of this process on a commercial scale is planned to be conducted in Shimane to study the effect of various parameters on the quality of liquid and gaseous products (ref: Pyrolytic oil from waste plastics, Yoshikawa laboratory, Tokyo Institute of Technlogy, at http://yoshikawalab.org/index.php?option=com_content&view=article&id=121&Itemid=200&lang=en).

Pyrolysis of mixed plastics: Mixed plastics with N_2-containing plastics produces the liquid fuel with nitrogen compounds, which in turn produces nitrogen oxide in the flue gas at combustion. Similarly, liquid fuel derived from waste plastics containing chlorine will cause corrosion to the reactor and burner and it will form hydrogen chloride and dioxins. Flue gas treatment is required to avoid the potential hazard of these chemicals.

4.7.3.4 Production of gaseous fuel

Gaseous products from pyrolysis of waste plastics are categorized into two major types: (i) syngas, a mixture of hydrogen and carbon monoxide, (ii) gaseous hydrocarbons such as methane and ethylene. Depending on gasification conditions, a mixture of hydrogen and methane will be obtained. The gas composition depends on reactor temperature, residence time of decomposing species during gasification process and other process conditions, decided by the reactor design.

Syngas, a mixture of hydrogen and carbon monoxide, was originally used for the production of methanol or ammonia. Syngas is also used as gaseous fuel for generating electricity.

There is another type of gasification for the production of gaseous hydrocarbon which is at research-to-demonstration stage (Kodera 2012). Thermal treatment of PE and PP at about 600°C or above with the residence time around 20 min gives mainly gaseous hydrocarbons. The initial decomposition product of plastics is liquid hydrocarbon, and further heat transfer to the vaporized portions results in the conversion of vaporized hydrocarbons into gaseous hydrocarbons of methane, ethylene, ethane, propylene, propane and the other gases with the formation of liquid hydrocarbons at about 10 to 20%.

4.7.4 Commercial systems

In Japan, under the 'Containers and Packaging Law', *mixed* plastics were converted into fuel oil through pyrolysis. Two plastic liquefaction facilities have been operating commercially in Japan: Niigata Plastic liquefaction Center (6000 TPA) and Sopporo Waste Plastics liquefaction Plant (14800 TPA) for several years. After initial obstacles, the Sopporo plant has maintained high levels of safety, stability and productivity (Fukushima et al. 2009). These commercial operations were later shut down due to the higher cost (about 80 yen/kg) compared to other treatment costs like that of cokes-oven treatment (about 40 yen/kg) (Kodera 2012).

The main pyrolysis units and technologies on an industrial scale include PYRO-PLEQ (rotary drum), Akzo (circulating FB), NRC (melt furnace), ConTherm technology (rotary drum), PKA pyrolysis (rotary drum), PyroMelt (melt furnace), BP (circulating FB), BASF (furnace) and NKT (circulating FB). The details of these processes are found in the literature, for example, Al-Salem et al. (2010).

The commercial plants under operation are described in Buekens (2006), UNEP (2009), Kodera (2012) and so on. However, the current operational status of these Plants is unclear. In Japan, middle-sized plants of about 3 to 6 t/d capacity are commercially operated for pyrolysis of industrial wastes since both the large-scale plants (20–40 t/d) and small-scale plants (1–1.5 t/d) are found to be costly to run (Kodera 2012).

In the U.S. there are several pyrolysis processes under pilot-scale demonstration: Agilyx, Global Climax Energy, Envion, Cynar, Covanta, JBI and others (GBB 2012). But, currently, many of them appear to be stalled. The author visited Agilyx plant in demonstration, located in Tigard, Oregon, and their process is briefly described.

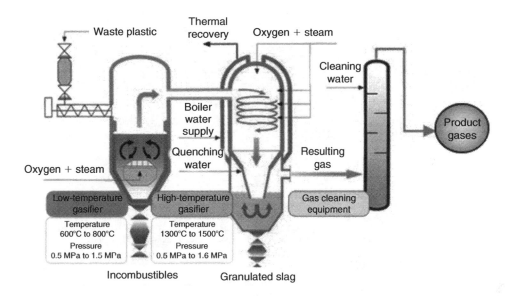

Figure 4.26 Schematic of FB gasifier with high temperature slagging furnace (PTFIG) (Source: http://www.eep.ebara.com/en/products/gas.html; reproduced with the permission of Ebara Environmental Plant Co. Ltd., Yasuhiko Hara).

EBARA's pressurized two-stage gasification system

The pressurized twin internally circulating fluidized-bed gasification system (PTIFG), which combines a low-temperature fluidized-bed gasification furnace with a subsequent high-temperature slagging gasification furnace is capable of generating syngas (H_2 and CO) from the plastic wastes, which serves as the feedstock for the ammonia synthesis process. This technology was jointly developed by Ube Industries, Ltd. and Ebara (Fig. 4.26).

The FB enables rapid gasification of rather heterogenous materials. The high temperature gasifier is a cyclone to collect fine ash particles on the wall. After vitrificatin the slag is discharged through the water seal. Both reactors operate at elevated pressure, around 8 bar. As the gasificaton takes place at high temperature, there is no risk of forming dioxins.

The operation to recycle plastic wastes in accordance with the Japanese Law was started in December 2000 as Ube-EBARA project with a capacity of 30 TPD. The technology can be directly applied to co-generation and to liquid fuel production, including methanol and can thus make a significant contribution to CO2 reduction and for automobile fuel. The syngas generated from wastes was intended to supply to the existing ammonia synthesis plant of Ube Ammonia for use as feedstock for ammonia synthesis (Fujimura et al. 2001). Another facility at Showa Denko K.K. operating from 2003 with throughput of 195 t/d derives hydrogen gas from waste plastics and utilizes it as raw material for ammonia synthesis (Ebara website).

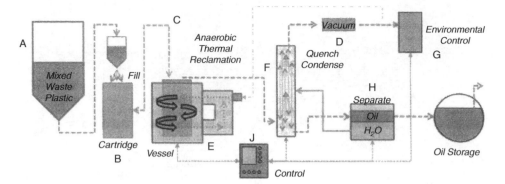

A: prepared mixed waste plastic; **B**: plastic filled into removable catridge, Direct contact catridge (DCC); **C**: DCC is placed into Plastic Reclamation Unit (PRU); **D**: System placed uder vacuum; **E**: PRU heated and heat is recirculated; **F**: gases move into Chromatographic Condenser separating process; **G**: non-condensable gases destructed for emission reduction and for heat usage; **H**: Oil is filtered, dried, cleaned, conditioned and transferred to storage tank; **J**: Integrated control system

Figure 4.27   Schematic of Agilyx Process diagram (Source: Company's Paper 2012; personal discussion with Ross Patten, Chairman and CEO).

Agilyx's plastics-to-crude oil technology

The Plant utilizes used-plastics of different types as feedstock and converts into synthetic crude oil through pyrolysis. The plasic feedstocks are either too difficult or contaminated to recycle by traditional recycling methods. Currently, the feedstock comes from local markets.

Process: The shredded and densified plastic feedstock is placed in a large, airtight processing vessel heated with an industrial burner, depolymerizing the plastic transforming from solid to a liquid to a gas. The gases are pulled from the catridge through a combination of temperature and vacuum into a condensing system where the gases are cooled and condensed into synthetic crude oil. The impurities in the feedstock and the fraction of non-convertibles result into a 'char' which is separated; the oil is filtered, dried, cleaned and conditioned before transferring to a tank which is sent to a refinery. An integrated control system controls the entire process with automation.

The non-condensable hydrocarbon gases are combusted leaving traces of air pollution other than water vapor and carbon dioxide (Fig. 4.27).

The solid char, around 7% of the original feedstock mass, can be used in cement factories or steel mills as an ignition fuel.

For this process, the feedstock should be appropriate which means less PET and PVC, and more of LDPE and HDPE.

One of the R&D efforts of Agilyx is to capture the non-condensable gases to supplement the energy required for the Process. This may lead to the reduction of natural gas use for heating and the process emissions. The process emissions are given in the Table 4.17 (Company white paper, Feb. 2012).

Agilyx Corp. is putting up 6th generation plastic-to-oil manufacturing plant (Marcus Hook Project) adjacent to Monroe Energy Refinery near Philadelphia, PA,

Table 4.17 Process emissions.

Major constituent	GHG Emissions factor (lb CO_2/lb pure constituent gas	Percent composition
n-Pentane	0.832	35.07
n-Butane	0.826	8.43
Isobutane	0.855	22.38
2-Methylpentane	0.835	6.06
2-Methyl-1-pentane	0.840	17.15
$C_5 H_{10}$ compound	0.857	10.91

Table 4.18 Fuel properties of oils derived from the pyrolysis of various wastes.

Property	Polyethylene[1]	Nylon[1]	Polyester[2] Styrene co-polymer	Tyre[3]	Bio-oil[4]	Gasoil[1]
Flash point (°C)	33.6	34.8	26.0	20	48.0	75
Pour point (°C)	2.7	−28	–	–	−30	−30
Ash (wt %)	0.013	0.018	0.53	0.002	0.003	0.01
Viscosity (cst 50/60°C)	2.19	1.8	3.9	2.38	9.726	1.3
Density (kg/m^3)	0.858	0.926	0.83	0.91	1.21	0.78
Carbon (wt %)	–	–	86.1	88.0	42.6	87.1
Hydrogen (wt %)	–	–	7.2	9.4	5.83	12.6
Sulphur (wt %)	0.01	0.01	0.0	1.45	0.01	0.2
Oxygen (wt %)	–	–	–	0.5	46–51	0.2
Initial B.Pt (°C)	–	–	75	100	–	180
50% B.Pt (°C)	–	–	189	264	–	300
CV (MJ/Kg)	52.3	44.4	33.6	42.1	17.6	46.0

1) Williams et al. 2006; 2) Williams 2003; 3) Cunliffe & Williams 2003; 4) Dynamotive Energy Systems Corp.

which is expected to go into operation from the end of the year 2015. This plant will source 50TPD of mixed plastic waste from local markets.

4.7.5   Fuel properties of pyrolytic oils

The plastics and tire pyrolysis oils were produced by conventional pyrolysis and the bio-oil by fast pyrolysis. The oils produced from plastics and tires have high calorific value, comparable to that of gas oil derived from petroleum (Table 4.18).

The bio-oil has a lower calorific value and contains >40 wt% oxygen due to the high oxygen content of the feed-stock biomass waste. The oxygen content is comprised of moisture (>20 wt%), and oxygenated compounds such as organic acids (5 to 10 wt%), aldehydes and hydroxyaldehydes (5 to 20 wt%), ketones and hydroxylketones (0–10 wt%) and phenolic compounds (15 to 30 wt%). Due to the high oxygen and moisture content in bio-oil, upgrading and refining is required for use as liquid transport fuels or chemical feedstock. Upgrading methods for bio-oils mainly include catalytic cracking, catalytic esterification and catalytic hydroprocessing (Williams 2012).

REFERENCES

Abbas-Abadi, M.S., Haghighi, M.N. & Yeganeh, H. (2012) The effect of temperature, catalyst, different carrier gases and stirrer on the produced transportation hydrocarbons of LLDPE degradation in a stirred reactor. *Journal of Analytical and Applied Pyrolysis*, 95, 198–204.

Achilias, D.S. (2007) Chemical recycling of poly (methyl methacrylate) by pyrolysis. Potential use of the liquid fraction as a raw material for the reproduction of the polymer. *European Polymer Journal*, 43(6), 2564–2575.

Achilias, D.S., Megalokonomos, P. & Karayannidis, G.P. (2006) Current trends in chemical recycling of polyolefins. *Journal of Environmental Protection and Ecology*, 7 (2), 407–413.

Achilias, D.S., Antonakou, E.V., Roupakias, C., Megalokonomos, P. & Lappas, A.A. (2008) Recycling techniques of polyolefins from plastic wastes. *Global NEST Journal*, 10 (1), 114–122.

Achilias, D.S., Andriotis, L., Koutsidis, I.A., Louka, D.A., Nianias, N.P., Siafaka, P., Tsagkalias, I. & Tsintzou, G. (2012) Recent advances in the chemical recycling of polymers (PP, PS, LDPE, HDPE, PVC, PC, Nylon, PMMA). In: Achilias, D.S. (ed.) *Material Recycling – Trends and Perspectives*. InTech. ISBN: 978-953-51-0327-1. Available from: http://www.intechopen.com/books/material-recycling-trends-and-perspectives/recent-advances-in-thechemical-recycling-of-polymers.

Adrados, A., de Marco, I., Caballero, B.M., López, A., Laresgoiti, M.F. & Torres, A. (2012) Pyrolysis of plastic packaging waste: A comparison of plastic residuals from material recovery facilities with simulated plastic waste. *Waste Management*, 32 (5), 826–832.

Adrados, A., de Marco, I., Lopez, A., Caballero, B.M. & Laresgoiti, M.F. (2013) Pyrolysis behaviour of different type of materials contained in the rejects of packaging waste sorting plants. *Waste Management*, 33 (1), 52–59.

AES (2004) *Investigation into Municipal Solid Waste Gasification for Power Generation, Advanced Energy Strategies*.

Agarwal, M. (August 2014) *An Investigation on the Pyrolysis of MSW*. Thesis submitted for Doctoral degree. Melbourne, School of Applied Sciences, RMIT University.

Aguado, J., Serrano, D.P., Escola, J.M., Garagorri, E. & Fernández, J.A. (2000) Catalytic conversion of polyolefins into fuels over zeolite beta. *Polymer Degradation and Stability*, 69 (1), 11–16.

Aguado, J., Serrano, D.P., Escola, J.M. & Garagorri, E. (2002a) Catalytic conversion of low-density polyethylene using a continuous screw kiln reactor. *Catalysis Today*, 75 (1–4), 257–262.

Aguado, R., Olazar, M., Gaisan, B., Prieto, R. & Bilbao, J. (2002b) Kinetic study of polyolefin pyrolysis in a conical spouted bed reactor. *Industrial & Engineering Chemistry Research*, 41 (18), 4559–4566.

Aguado, J., Serrano, D.P. & Escola, J.M. (2006) Catalytic upgrading of plastic wastes. In: Scheirs, J. & Kaminsky, W. (eds.) *Feedstock Recycling and Pyrolysis of Waste Plastics*. Mostoles, Spain, John Wiley & Sons, Ltd. p. 73110.

Aguado, J., Serrano, D.P., Vicente, G. & Sanchez, N. (2007a) Enhanced production of alpha-olefins by thermal degradation of High-Density Polyethylene (HDPE) in decalin solvent: Effect of the reaction time and temperature. *Industrial & Engineering Chemistry Research*, 46 (11), 3497–3504.

Aguado, J., Serrano, D.P., Miguel, G.S., Escola, J.M. & Rodriguez, J.M. (2007b) Catalytic activity of zeolitic and mesostructured catalysts in the cracking of pure and waste polyolefins. *Journal of Analytical and Applied Pyrolysis*, 78, 153–161.

Aguado, J., Serrano, D.P. & Escola, J.M. (2008) Fuels from waste plastics by thermal and catalytic processes: A review. *Industrial & Engineering Chemistry Research*, 47 (21), 7982–7992.

Ahrenfeldt, J. (2007) *Characterisation of Biomass Producer Gas as Fuel for Stationary Gas Engines in CHP Production*. Doctoral Thesis. Lyngby, Technical University of Denmark.

Al-Salem, S.M., Lettieri, P. & Baeyens, J. (2009) Recycling and recovery routes of plastic solid waste (PSW): A review. *Waste Management*, 29, 2625–2643.

Al-Salem, S.M., Lettieri, P. & Baeyens, J. (2010) The valorization of plastic solid waste (PSW) by primary to quarternary routes: From reuse to energy and chemicals. *Progress in Energy and Combustion Science*, 36, 103–129.

American Chemistry Council. *Lifecycle of a Plastic Product*. Available from: http://plastics.americanchemistry.com/Life-Cycle.

Anna, P., Yang, W. & Lucas, C. (2006) Development of a thermally homogeneous gasifier system using high-temperature agents. *Clean Air*, 7 (4), 363–379.

An'shakov, A.S., Faleev, V.A., Danilenko, A.A., Urbakh, E.K. & Urbakh, A.E. (2007) Investigation of plasma gasification of carbonaceous technogeneous wastes. *Thermophysics and Aeromechanics*, 14, 607–616.

Arabiourrutia, M., Elordi, G., Lopez, G., Borsella, E., Bilbao, J. & Olazar, M. (2012) Characterization of the waxes obtained by the pyrolysis of polyolefin plastics in a conical spouted bed reactor. *Journal of Analytical and Applied Pyrolysis*, 94, 230–237.

Arena, U. (2012) Process and technological aspects of municipal solid waste gasification—A review. *Waste Management*, 32, 625–639.

Arena, U. & Mastellone, M.L. (2006) Fluidized bed pyrolysis of plastic wastes. In: Scheirs, J. & Kaminsky, W. (eds.) *Feedstock Recycling and Pyrolysis of Plastic Wastes: Converting Waste Plastics into Diesel and Other Fuels*. Mostoles, Spain, John Wiley & Sons.

Arena, U., Zaccariello, L. & Mastellone, M.L. (2010) Fluidized bed gasification of waste-derived fuels. *Waste Management*, 30, 1212–1219.

Arena, U., Di Gregorio, F., Amorese, C. & Mastellone, M.L. (2011) A techno-economic comparison of fluidized bed gasification of two mixed plastic wastes. *Waste Management*, 31, 1494–1504.

Assumpção, L.C.F.N, Carbonell, M.M. & Marques, M.R.C. (2011) Co-pyrolysis of polypropylene waste with Brazilian heavy oil. *Journal of Environmental Science and Health, Part A*, 46, 461–464.

Ates, F., Miskolczi, N. & Borsodi, N. (2013) Comparison of real waste (MSW and MPW) pyrolysis in batch reactor over different catalysts, Part I: Product yields, gas and pyrolysis oil properties. *Bioresource Technology*, 133, 443–454.

Belgiorno, V., De Feo, G., Della Rocca, C. & Napoli, R.M.A. (2003) Energy from gasification of solid wastes. *Waste Management*, 23, 1–15.

Bendix, D. & Hebecker, D. (2003) Energy recovery from waste and plasma conversion. *High Temperature Material Processes*, 7, 435–454.

Bhaskar, T., Azhar Uddin, Md., Murai, K., Kaneko, J., Hamano, K., Kusaba, T., Muto, A. & Sakata, Y. (2003) Comparison of thermal degradation products from real municipal waste plastic and model mixed plastics. *Journal of Analytical and Applied Pyrolysis*, 70, 579–587.

Bhaskar, T., Tanabe, M., Muto, A. & Sakata, Y. (2006) Pyrolysis study of a PVDC and HIPS-Br containing mixed waste plastic stream: Effect of the poly (ethylene terephthalate). *Journal of Analytical and Applied Pyrolysis*, 77, 68–74.

Bhaskar, T., Hall, W.J., Mitan, N.M.M., Muto, A., Williams, P.T. & Sakata, Y. (2007) Controlled pyrolysis of polyethylene/polypropylene/polystyrene mixed plastics with high impact polystyrene containing flame retardant: Effect of decabromo diphenylethane (DDE). *Polymer Degradation and Stability*, 92, 211–221.

Bi, X.T.T. & Liu, X.H. (2010) High density and high solids flux CFB risers for steam gasification of solids fuels. *Fuel Processing Technology*, 91 (8), 915–920.

Biganzoli, L., Ilyas, A., van Praagh, M., Persson, K.M. & Grosso, M. (2013) Aluminium recovery Vs hydrogen production as resource recovery options for fine MSWI bottom ash fraction. *Waste Management*, 33, 1174–1181.

Blazso, M. & Jakab, E. (1999) Effect of metals, metal oxides, and carboxylates on the thermal decomposition processes of poly (vinyl chloride). *Journal of Analytical and Applied Pyrolysis*, 49, 125–143.

Bolhar-Nordenkampf, J.P., Hofbauer, T. & Aichernig, H.C. (2006) Analysis of thermoelectric generators replacing low temperature heat exchangers in the biomass CHP plant Gussing. In: *Proc. of the 8th Biennial Conference on Engineering Systems Design and Analysis*. Vol. 4. pp. 305–313.

Bosmans, A., Vanderreydt, I., Geysen, D. & Helsen, L. (2013a) The crucial role of Waste-to-Energy technologies in enhanced landfill mining: A technology review. *Journal of Cleaner Production*, 55, 10–23.

Bosmans, A., Wasan, S. & Helsen, L. (2013b) Waste to clean syngas: Avoiding tar problems. In: *2nd Intl. Symposium on Enhanced Landfill Mining, Houthalen-Helchter, 14-16/10/2013*.

Boulos, M.I. (1996) New frontiers in thermal plasma processing. *Pure and Applied Chemistry*, 68 (5), 1007–1010.

Bowyer, J. & Fernholz, K. (2009) *Plasma Gasification: An Examination of the Health Safety and Environmental Records of Established Facilities*. Dovetail Partners, Inc. Available from: http://www.dovetailinc.org/files/u1/PlasmaGasificationPresentation.pdf.

Boyle, G. (2004) *Renewable Energy Power for a Sustainable Future*. Oxford, UK, Oxford University Press.

Brage, C., Qizhuang, U. & Sjostrom, K. (1997) Use of amino phase adsorbent for biomass tar sampling and separation. *Fuel*, 76 (2), 137–142.

Brebu, M., Bhaskar, T., Murai, K., Muto, A., Sakata, Y. & Azhar Uddin, Md. (2004) Thermal degradation of PE and PS mixed with ABS-Br and debromination of pyrolysis oil by Fe- and Ca-based catalysts. *Polymer Degradation and Stability*, 84, 459–467.

BREF Report (December 2001) Draft of a German report with basic information for a BREF-Document "Waste Incineration". BREF Report English—FTP Direct Listing. Available from: files.gamta.It/aaa/Tipk/tipk/4_kiti%20GPGB/63.pdf.

Brems, A., Baeyens, J., Beerlandt, J. & Dewil, R. (2011) Thermogravimetric pyrolysis of waste polyethylene-terephthalate and polystyrene: A critical assessment of kinetics modeling. *Resources, Conservation and Recycling*, 55, 772–781.

Brems, A., Dewil, R., Baeyens, J. & Zhang, R. (2013) Gasification of plastic waste as waste-to-energy or waste-to-syngas recovery route. *Natural Science*, 5, 695–704 (Open access).

Brickner, B. (2012) *WtE and Alternative Conversion Technologies—Experiences and Opportunities*. MWMA 2012 Fall Summit, September 12, 2012. Available from: www.gbbinc.com/speaker/BricknerMWMA2012.pdf.

Bridgwater, A.V. (1995) The technical and economic-feasibility of biomass gasification for power-generation. *Fuel*, 74 (5), 631–653.

Bridgwater, A.V. (2003) Renewable fuels and chemicals by thermal processing of biomass. *Chemical Engineering Journal*, 91, 87–102.

Buah, W.K., Cunliffe, A.M. & Williams, P.T. (2007) Characterization of products from the pyrolysis of municipal solid waste. *Process Safety & Environment Protection*, 85, 450–457.

Buekens, A. (2006) Introduction to feedstock recycling of plastics. In: Scheirs, J. & Kaminsky, W. (eds.) *Feedstock Recycling and Pyrolysis of Waste Plastics*. New York, NY, Wiley.

Butler, E., Devlin, G. & McDonnell, K. (2011) Waste polyolefins to liquid fuels via pyrolysis: Review. *Waste and Biomass Valorization*, 2, 227–255.

Byun, Y., Namkung, W., Cho, M., Chung, J.W., Kim, Y.S., Lee, J.H., Lee, C.R. & Hwang, S.M. (2010) Demonstration of thermal plasma gasification/vitrification for municipal solid waste treatment. *Environmental Science & Technology*, 44, 6680–6684.

Byun, Y., Cho, M., Chung, J.W., Namkung, W., Lee, H.D., Jang, S.D., Kim, Y.-S., Lee, J.H., Lee, C.R. & Hwang, S.M. (2011) Hydrogen recovery from the thermal plasma gasification of solid waste. *Journal of Hazardous Materials*, 190, 317–323.

Byun, Y., Cho, M., Hwang, S. & Chung, J. (2013) Thermal plasma gasification of municipal solid waste (MSW), Chapter 7. In: Gasification for Practical Applications. INTECH. pp. 183–208.

Calaminus, B. & Stahlberg, R. (1998) Continuous in-line gasification/vitrification process for thermal waste treatment: Process technology and current status of projects. *Waste Management*, 18, 547–556.

Camacho, S.L. (1988) Industrial worthy plasma torches: State-of-the-art. *Pure and Applied Chemistry*, 60, 619–632.

Campbell, F. (2008) *An Overview of the History and Capabilities of the ThermoSelect Technology*. Presented at Federation of New York Solid Waste Associations Solid Waste and Recycling Conference, November 12, 2008.

Caponero, J., Tenorio, J.A.S., Levendis, Y.A. & Carlson, J.B. (2005) Emissions of batch combustion of waste tire chips: The pyrolysis effect. *Combustion Science and Technology*, 177 (2), 347–381.

Castaldi, M.J. (November 12, 2008) *Principles and Essential Design Characteristics of Gasification and Anaerobic Digestion Systems for Solid Waste Processing, Waste Conversion Technologies: Theory and Practice*. SWANA NYS Chapter.

Castaldi, M.J. & Themelis, N.J. (2010) The case for increasing the global capacity for waste to energy (WTE). *Waste and Biomass Valorization*, 1, 91–105.

CH2M Hill (May 2009) *WtE Review of Alternatives*. Final report prepared for Regional District of North Okanagan by CH2M Hill Canada Ltd.

Chang, J.S., Gu, B.W., Looy, P.C., Chu, F.Y. & Simpson, C.J. (1996) Thermal plasma pyrolysis of used old tires for production of syngas. *Journal of Environmental Science and Health, Part A*, 31 (7), 1781–1799.

Chapman, C.D., Williams, J.K., Heanley, C.P., Iddles, D.M., Forde, A.J. & Risdon, G. (1995) Development of a plasma reactor for the treatment of incinerator ashes. In: *Proc. Intl. Symp. on Environmental Technologies: Plasma Systems and Applications*. Atlanta, GA, Georgia Tech Research Corp. pp. 129–140.

Chapman, C.D., Deegan, D.E. & Ly, H. (2007) Integrated process solution for advanced thermal treatment of hazardous waste materials using plasma arc technology. In: *Proc. Intl. Round Table on Thermal Plasma Fundamentals and Applications, Sharm el Sheikh, Egypt*.

Chen, D., Yin, L., Wang, H. & He, P. (2014) Pyrolysis techniques for MSW: A review. *Waste Management*, 34, 2466–2486.

Cheng, T.W., Chu, J.P., Tzeng, C.C. & Chen, Y.S. (2002) Treatment and recycling of incinerated ash using thermal plasma technology. *Waste Management*, 22, 485–490.

Cherednichenko, V.S., Anshakov, A.S., Danilenko, A.A., Michajlov, V.E., Faleev, V.A. & Kezevich, D.D. (2002) Domestic waste plasma gasification technology and its comparison with ordinary one burning on the final products. *Ecology, Electrotechnology and Waste Process—KORUS 2002*, 211–213.

Chopra, S. & Jain, A.K. (April 2007) A review of fixed bed gasification systems for biomass. *Agricultural Engineering International: The CIGR Ejournal*. Invited Overview 5: IX.

Chu, J.P., Hwang, I.J., Tzeng, C.C., Kuo, Y.Y. & Yu, Y.J. (1998) Characterization of vitrified slag from mixed medical waste surrogates treated by a thermal plasma system. *Journal of Hazardous Materials*, 58, 179–194.

Ciferno, J.P. & Marano, J.J. (2002) *Benchmarking Biomass Gasification Technologies for Fuels, Chemicals, and Hydrogen Production*. E2S for National Energy Technology Laboratory.

Ciliz, N.K. Ekinci, E. & Snape, C.E. (2004) Pyrolysis of virgin and waste polypropylene and its mixtures with waste polyethylene and polystyrene. *Waste Management*, 24, 173–181.

Circeo, L. (2008) *Plasma Arc Gasification of MSW*. Available from: http://www.energy.ca.gov/proceedings/2008-ALT-1/documents/2009-02-17_workshop/presentations/Louis_Circeo-Georgia_Tech_Research_Institute.pdf.

Cit, I., Smag, A., Yumak, T., Ucar, S., Misirlioglu, Z. & Canel, M. (2010) Comparative pyrolysis of polyolefins (PP and LDPE) and PET. *Polymer Bulletin*, 64, 817–834.

Climatetechwiki. *Gasification of Municipal Solid Waste for Large-Scale Electricity/Heat*. Available from: www.climatetechwiki.org/technology/msw.

Colby, S., Turner, Z., Utley, D. & Duy, C. (2006) Treatment of radioactive reactive mixed waste. In: *WM'06 Conference, Feb. 26–March 2, 2006, Tucson, AZ*.

Colpan, C.O., Hamdullahpur, F., Dincer, I. & Yoo, Y. (2010) Effect of gasification agent on the performance of solid oxide fuel cell and biomass gasification systems. *International Journal of Hydrogen Energy*, 35 (10), 5001–5009.

Conesa, J.A., Font, R., Marcilla, A. & Garcia, A.N. (1994) Pyrolysis of polyethylene in a fluidized bed reactor. *Energy & Fuels*, 8 (6), 1238–1246.

Conssoni, S. & Vigano, F. (2012) Waste gasification vs conventional WTE: A comparison. *Waste Management*, 32 (4), 653–666.

Costa, P.A., Pinto, F.J., Ramos, A.M., Gulyurtlu, I.K., Cabrita, I.A. & Bernardo, M.S. (2007) Kinetic evaluation of the pyrolysis of polyethylene waste. *Energy & Fuels*, 21 (5), 2489–2498.

Croezen, H. & Sas, H. (1997) *Evaluation of the Texaco Gasification Process for Treatment of Mixed Household Waste*. Final Report of Phase 1 and 2. CE, Delft, The Netherlands.

Cunliffe, A.M. & Williams, P.T. (2003) Characterisation of waste by pyrolysis. *Fuel*, 82, 2223–2230.

De Filippis, P., Borgianni, C., Paolucci, M. & Pochetti, F. (2004) Prediction of syngas quality for two-stage gasification of selected waste feedstocks. *Waste Management*, 24 (6), 633–639.

DEFRA (February 2013) *Incineration of Municipal Solid Waste*. Govt. of UK. Available from: www.defra.gov.uk/publications/.

De Souza-Santos, M.L. (2004) *Solid Fuels Combustion and Gasification*. New York, NY, Marcel Dekker Inc. ISBN: 0-8247-0971-3.

Di Gregorio, F. & Zaccariello, L. (2012) Fluidized bed gasification of a packaging derived fuel: Energetic, environmental and economic performances comparison for waste-to-energy plants. *Energy*, 42, 331–341.

Deegan, D., Chapman, C. & Bowen, C. (2002) The production of shaped glass-ceramic materials from inorganic waste precursors using controlled atmospheric DC plasma vitrification and crystallization. In: *Proc. 2002 EMRS Conference, Strasbourg, France*.

Delegation of the Panel on Environmental Affairs (2014) Report on the duty visit to the United Kingdom, the Netherlands, Denmark and Sweden to study these countries' experience on thermal waste treatment facilities, LC Paper No. CB (1)359/14-15, Legislative Council of the Hong Kong Special Administrative Region, 2–8 March 2014.

Dogru, M., Howarth, C.R., Akay, G., Keskinler, B. & Malik, A.A. (2002) Gasification of hazelnut shells in a downdraft gasifier. *Energy*, 27 (5), 415–427.

Ducharme, C. (September 2010) *Technical and Economic Analysis of Plasma-Assisted Waste-to-Energy Processes*. Thesis submitted for MS degree in Earth Resources Engineering. New York, NY, Department of Earth and Environmental Engineering, Columbia University.

Dvirka and Bartilucci Consulting Engineers (2007) *Waste Conversion Technologies: Emergence of a New Option or the Same Old Story?*

Dynamotive Energy Systems Corp., Richmond, Canada. Available from: http://www.dynamotive.com/assets/resources/PDF/PIB-BioOil.pdf.

E4tech (2009) *Review of Technologies for Gasification of Biomass and Wastes*. E4tech June 20bara09. Available from: http://www.nnfcc.co.uk/tools/review-of-technologies-for-gasification-of-biomass-and -wastes-nnfcc-09-008.

EBARA Environmental Plant Co. *Fluidized-Bed Gasification Technologies*. Available from: www.eep.ebara.com/en/products/gas.html [Retrieved 17th June, 2015].

Eden, R. *Medical Pyrolysis Plant Resurrecting Avonmouth*. Waste Management World. Available from: www.waste-management-world.com/.../medic.

Elordi, G., Olazar, M., Aguado, R., Lopez, G., Arabiourrutia, M. & Bilbao, J. (2007) Catalytic pyrolysis of high density polyethylene in a conical spouted bed reactor. *Journal of Analytical and Applied Pyrolysis*, 79 (1–2), 450–455.

Elordi, G., Olazar, M., Lopez, G., Amutio, M., Artetxe, M., Aguado, R. & Bilbao, J. (2009) Catalytic pyrolysis of HDPE in continuous mode over zeolite catalysts in a conical spouted bed reactor. *Journal of Analytical and Applied Pyrolysis*, 85 (1–2), 345–351.

ENERGOS-2 *The Process*. Available from: http://www.energ.co.uk/energy-from-waste-process.

Energy Recovery Council (2010) *The 2010 ERC Directory of Waste-to-Energy Plants*.

Ergut, A., Levendis, Y.A. & Carlson, J. (2007) Emissions from the combustion of polystyrene, styrene and ethylbenzene under diverse conditions. *Fuel*, 86 (12–13), 1789–1799.

ESTET (March 2014) *The Viability of Advanced Thermal Treatment of MSW in the UK*. Prepared by Fichtner Consulting Engineers Ltd, published by ESTET, London, UK.

Faaij, A., Van Ree, R., Waldheim, L., Olsson, E., Oudhuis, A., Van Wijk, A., Daey-Ouwens, C. & Turkenburg, W. (1997) Gasification of biomass wastes and residues for electricity production. *Biomass and Bioenergy*, 12 (6), 387–407.

Fabry, F., Rehmet, C., Rohani, V. & Fulcheri, L. (2013) Waste gasification by thermal plasma: A review. *Waste and Biomass Valorization*, 2013, 4 (3), 421–439.

Faravelli, T., Bozzano, G., Scassa, C., Perego, M., Fabini, S., Ranzi, E. & Dente, M.J. (1999) Gas product distribution from polyethylene pyrolysis. *Journal of Analytical and Applied Pyrolysis*, 52 (1), 87–103.

Fauchais, P. (2007) *Technologies plasma: Applications au traitement des déchets*. Techniques de l'Ingénieur G2055:1-11.

Feasibility Study of Thermal Waste Treatment/Recovery Options in the Limerick/Clare/Kerry Region; MDE0267Rp0003 32 Rev F01; 4. Thermal Waste Treatment: Technologies.

Ficktner, Babtie (2008) *States of Jersey—Solid Waste Strategy—Technology Review 2008*, Author: Weatherby, J.

Fonts, I., Azuara, M., Gea, G. & Murillo, M.B. (2009) Study of the pyrolysis liquids obtained from different sewage sludge. *Journal of Analytical and Applied Pyrolysis*, 85, 184–191.

Fonts, I., Gea, G., Azuara, M., Abrego, J. & Arauzo, J. (2012) Sewage sludge pyrolysis for liquid production: A review. *Renewable and Sustainable Energy Reviews*, 16, 2781–2805.

Foth Infrastructure & Environment, LLC (July 2013) *Alternative Technologies for Municipal Solid Waste*. Report prepared for Ramsey Washington County Resource Recovery Project.

Friberg, R. & Blasiak, W. (2002) Measurements of mass flux and stoichiometry of conversion gas from three different wood fuels as function of volume flux of primary air in packed-bed combustion. *Biomass and Bioenergy*, 23 (3), 189–208.

Fujimura, H., Oshita, T. & Naruse, K. (2001) *Fluidized Bed Gasification and Slagging Combustion System*. Presented at the IT3 Conference, May 14–18, 2001, Philadelphia, USA. e-mail:naruse@shi.ebara.co.jp.

Fukushima, M., Shioya, M., Wakai, K. & Ibe, H. (2009) Toward maximizing the recycling rate in Sapporo waste plastics liquefaction plant. *Journal of Material Cycles and Waste Management*, 11, 11–18.

Fushimi, C., Araki, K., Yamaguchi, Y. & Tsutsumi, A. (2003) Effect of heating rate on steam gasification of biomass. 1. Reactivity of char. *Industrial & Engineering Chemistry Research*, 42 (17), 3922–3928.

Fyffe, J., Breckel, A.C., Townsend, A.K. & Webber, M.E. (2012) *Residue-Derived Solid Recovery Fuel for Use in Cement Kilns*. Available from: http://plastics.americanchemis try.com/Sustainability-Recycling/Energy-Recovery/Residue-Derived-Solid-Recovered-Fuel-for-Use-in-Cement-Kilns.pdf.

Galeno, G., Minutillo, M. & Perna, A. (2011) From waste to electricity through integrated plasma gasification/fuel cell (IPGFC) system. *International Journal of Hydrogen Energy*, 36, 1692–1701.

Gao, F. (2010) *Pyrolysis of Waste Plastics into Fuels*. Doctoral Thesis in Chemical and Process Engineering. New Zealand, University of Canterbury.

Garcia, A.N., Font, R. & Marcilla, A. (1995) Gas production by pyrolysis of MSW at high temperature in a fluidized bed reactor. *Energy Fuels*, 9, 648–658.

GBB (Gershman, Brickner & Bratton, Inc.) (2013) *Gasification of Non-Recycled Plastics from Municipal Solid Waste in the United States*. Prepared for 'The American Chemistry Council,' August 13, 2013. Available from: http://plastics.americanchemistry.com/Sustainability-Recycling/Energy-Recovery/Gasification-of-Non-Recycled-Plastics-from-Municipal-Solid-Waste-in-the-United-States.pdf.

Gendebien, A., Leavens, A., Blackmore, K., Godley, A., Lewin, K., Whiting, K.J., Davis, R., Giegrich, J., Fehrenback, H., Gromke, U., del Bufalo, N. & Hogg, D. (2003) *Refuse Derived Fuel, Current Practice and Perspectives*. Tech. Rep. July, European Commission—Directorate General Environment.

Gershman, Brickner & Bratton, Inc. (2012). Waste-to-Energy and Alternative Conversion technologies – Experience & Opportunities, MWMA (Municipal Waste management association) 2012 Fall Summit, September 12, 2012.

Gómez-Barea, A. & Leckner, B. (2010) Modeling of biomass gasification in fluidized bed. *Progress in Energy and Combustion Science*, 36 (4), 444–509.

Gómez-Barea, A., Vilches, L.F., Leiva, C., Campoy, M. & Fernández-Pereira, C. (2009) Plant optimisation and ash recycling in fluidised bed waste gasification. *Chemical Engineering Journal*, 146 (2), 227–236.

Gonçalves, C.K., Tenorio, J.A.S., Levendis, Y.A. & Carlson, J.B. (2007a) Emissions from premixed combustion of polystyrene. *Energy & Fuels*, 22 (1), 354–362.

Gonçalves, C.K., Tenório, J.A.S., Levendis, Y.A. & Carlson, J.B. (2007b) Emissions from premixed combustion of gasified polyethylene. *Energy & Fuels*, 22 (1), 372–381.

Göransson, K., Söderlind, U., He, J. & Zhang, W. (2011) Review of syngas production via biomass DFBGs. *Renewable and Sustainable Energy Reviews*, 15, 482–492.

Gustafsson, E., Strand, M. & Sanati, M. (2007) Physical and chemical characterization of aerosol particles formed during the thermochemical conversion of wood pellets using a bubbling fluidized bed gasifier. *Energy & Fuels*, 21, 3660–3667.

He, M., Hu, Z., Xiao, B., Li, J., Guo, X., Luo, S., et al. (2009) Hydrogen-rich gas from catalytic steam gasification of municipal solid waste (MSW): Influence of catalyst and temperature on yield and product composition. *International Journal of Hydrogen Energy*, 34 (1), 195–203.

He, M., Xiao, B., Liu, S., Hu, Z., Guo, X., Luo, S. & Yang, F. (2010) Syngas production from pyrolysis of municipal solid waste (MSW) with dolomite as downstream catalysts. *Journal of Analytical and Applied Pyrolysis*, 87, 181–187.

Heberlein, J. & Murphy, A.B. (2008) Thermal plasma waste treatment. *Journal of Physics D: Applied Physics*, 41, 053001, Inst. of Physics, UK.

Helsen, L. & Bosmans, A. (2010) Waste-to-Energy through thermochemical processes: Matching waste with process. In: *1st Intl. Symp. on Landfill Mining, Houthalen-Helchteren, 4–6 October 2010*.

Higman, C. & Burgt, M.v.d. (2003) *Gasification*. Boston, MA, Elsevier/Gulf Professional Publishers, 391.

Higman, C. & Burgt, M.v.d. (2008) *Gasification*. 2nd edition. Burlington, MA, Elsevier Science. 1 v.

Horikawa, S., Takai, Y., Ukei, H., Azuma, N. & Ueno, A. (1999) Chlorine gas recovery from polyvinyl chloride. *Journal of Analytical and Applied Pyrolysis*, 51, 167–179.

Huang, H., Tang, L. & Wu, C.Z. (2003) Characterization of gaseous and solid product from thermal plasma pyrolysis of waste rubber. *Environmental Science & Technology*, 37, 4463–4467.

Hur, M. & Hong, S.H. (2002) Comparative analysis of turbulent effects on thermal plasma characteristics inside the plasma torches with rod- and well-type cathodes. *Journal of Physics D: Applied Physics*, 35, 1946–1954.

IEA Bioenergy. *Accomplishments from IEA Bioenergy Task 36: Integrating Energy Recovery into Solid Waste Management Systems (2007–2009)—End of Task Report*. Chapter 4: Beciden, M., SINTEF, Vehlow, J., KIT & Howes, P., AEA (contributors). Available from: www.ieabioenergytask36.org/.../2007-2009/Introduction_Final.pdf.

Inaba, T., Nagano, M. & Endo, M. (1999) Investigation of plasma treatment for hazardous wastes such as fly ash and asbestos. *Electrical Engineering in Japan*, 126, 73–82.

International Atomic Energy Agency (IAEA) (2006) *Application of Thermal Technologies for Processing of Radioactive Waste*.

Ishikawa, M., Terauchi, M., Komori, T. & Yasuraoka, J. (2008) *Development of High Efficiency Gas Turbine Combined Cycle Power Plant*. Mitsubishi Heavy Industries, Ltd. Technical Review. 45 (1), 15–17.

Islam, M.N., Beg, M.R.A. & Islam, M.R. (2005) Pyrolysis oil from fixed bed pyrolysis of MSW and its characterization. *Renewable Energy*, 30, 413–420.

ISWA (January 2013) *White Paper—Alternative Waste Conversion Technologies*. Prepared by ISWA Working group on Energy Recovery. Leading authors: Lamers, F., Fleck, E., Pelloni, L. & Kamuk, B. International Solid Waste Association.

Ivanova, S.R., Gumerova, E.F., Minsker, K.S., Zaikov, G.E. & Berlin, A.A. (1990) Selective catalytic degradation of polyolefins. *Progress in Polymer Science*, 15 (2), 193–215.

Jameel, H., Keshwani, D., Carter, S. & Treasure, T. (2010) Thermochemical conversion of biomass to power and fuels. In: Cheng, J. (ed.) *Biomass to Renewable Energy Processes*. Boca Raton, FL, CRC Press. pp. 447–487.

Jayarama Reddy, P. (2014) *Clean Coal Technologies for Power Generation*. The Netherlands, CRC Press/Balkema, Leiden. ISBN: 978-1-138-00020-9 (Hbk); ISBN: 978-0-203-76886-0 (eBook PDF).

Jenkins, S.D. (2007) *Conversion Technologies: A New Alternative for MSW Management*. Earthscan.

Jiao, Z., Yu-qi, J., Yong, C., Jun-ming, W., Xu-guang, J. & Ming-jiang, N. (2009) Pyrolysis characteristics of organic components of municipal solid waste at high heating rates. *Waste Management*, 29, 1089–1109.

Jimbo, H. (1996) Plasma melting and useful application of molten slag. *Waste Management*, 16, 417–422.

Juniper.com (2008) *The Alter NRG/Westinghouse Plasma Gasification Process: Independent Waste Technology Report*. In: Juniper.com, Bisley, England.

Juniper Consultancy Services Inc. (2009) *Nippon Steel Gasification Process Review*. Available from: http://www.juniper.co.uk/Publications/Nippon_steel.html.

Kaminsky, W. (1980) Pyrolysis of plastic waste and scrap tyres in a fluid bed reactor. *Resource Recovery and Conservation*, 5 (3), 205–216.

Kaminsky, W. (1992) Plastics, recycling. *Ullmann's Encyclopedia of Industrial Chemistry*, A21, 57–73.

Kaminsky, W. & Zorriqueta, I.-J.N. (2007) Catalytical and thermal pyrolysis of polyolefins. *Journal of Analytical and Applied Pyrolysis*, 79 (1–2), 368–374.

Kaminsky, W., Schlesselmann, B. & Simon, C. (1995) Olefins from polyolefins and mixed plastics by pyrolysis. *Journal of Analytical and Applied Pyrolysis*, 32 (Part 1), 19–27.

Kaminsky, W., Schlesselmann, B. & Simon, C. (1996) Thermal degradation of mixed plastic waste to aromatics and gas. *Polymer Degradation and Stability*, 53 (2), 189–197.

Kaminsky, W., Predel, M. & Sadiki, A. (2004) Feedstock recycling of polymers by pyrolysis in a fluidised bed. *Polymer Degradation and Stability*, 85 (3 special issue), 1045–1050.

Kamińska-Pietrzak, N. & Smoliński, A. (2013) Selected environmental aspects of gasification and co-gasification of various types of waste. *Journal of Sustainable Mining*, 12 (4), 6–13.

Katou, K., Asou, T., Kurauchi, Y. & Sameshima, R. (2001) Melting municipal solid waste incineration residue by plasma melting furnace with a graphite electrode. *Thin Solid Films*, 386, 183–188.

Khodakov, A.Y., Chu, W. & Fongarland, P. (2007) Advances in the development of novel cobalt Fischer–Tropsch catalysts for synthesis of long-chain hydrocarbons and clean fuels. *ChemInform*. doi:10.1002/chin.200733255.

Kim, H.Y. (2003) A low cost production of hydrogen from carbonaceous wastes. *International Journal of Hydrogen Energy*, 28, 1179–1186.

Klein, A. (2002) *Gasification: An Alternative Process for Energy Recovery and Disposal of Municipal Solid Wastes*. MS Thesis. New York, NY, Earth Engineering Center, Colombia University. Available from: http//www.seas.columbia.edu/earth/wtert/sofos/kelin_thesis.pdf.

Klein, A. & Themelis, N.J. (April 2003) Energy recovery from MSW by gasification. In: *North America Waste to Energy Conf. 11 (Nawtec-11) Proceedings, ASME International, Tampa, FL*. pp. 241–253.

Knoef, H. & Ahrenfeldt, J. (2005) *Handbook: Biomass Gasification*. Enschede, BTG Biomass Technology Group. 378 pp.

Kodera, Y. (2012) Plastic recycling—Technology and business in Japan, Chapter 10. In: Rebellon, L.F.M. (ed.) *Waste Management—An Integrated Vision*. InTech. ISBN: 978-953-51-0795-8, Chapters published October 26, 2012 under CC BY 3.0 license, doi:10.5772/3150.

Kolb, T. & Seifert, H. (2002) Thermal waste treatment: State of the art—A summary. In: *Waste Management 2002: The Future of Waste Management in Europe, 7–8 Oct. 2002, Strasbourg, France*. VDI GVC, Dusseldorf, Germany.

Kordylewski, W. (2005) *Spalanie i paliwa (Combustion and Fuel)*. Wrocław, Oficyna Wydaw. Politechniki Wrocławskiej.

Krasovskaya, L.I. & Mossé, A.L. (1997) Use of electric-arc plasma for radioactive waste immobilization. *Journal of Engineering Physics and Thermophysics*, 70, 631–638.

Ladwig, M., Lindvall, K. & Conzelman, R. (2007) The realised gas turbine process with sequential combustion: Experience, state of development, prospects. *VGB Powertech*, 87 (10), 30–35.

Larsson, R. (2014) *Energy Recovery of Metallic Aluminium in MSWI Bottom Ash, Different Approaches to Hydrogen Production from MSWI Bottom Ash: A Case Study*. MS Thesis. Sweden, Tekniska Hogskolan, Umea University.

Lawrence, A.R. (1998) Energy from MSW: A comparison with coal combustion technology. *Progress in Energy and Combustion Science*, 24 (6), 545–564.

Lee, J.W., Yu, T.U., Lee, J.W., Moon, J.H., Jeong, H.J., Park, S.S., Yang, W. & Lee, U.D. (2013) Gasification of mixed plastic wastes in a moving-grate gasifier and application of the producer gas to a power generation engine. *Energy & Fuels*, 27 (4), 2092–2098.

Lemmens, B., Elslander, H., Vanderrydt, I., Peys, K., Diels, L., Osterlinck, M. & Joos, M. (2007) Assessment of plasma gasificaton of high calorific value waste stream. *Waste Management*, 27, 1562–1569.

Li, A.M., Li, X.D., Li, S.Q., Ren, Y., Shang, N., Chi, Y., Yan, J.H. & Cen, K.F. (1999) Experimental studies on municipal solid waste pyrolysis in a laboratory-scale rotary kiln. *Energy*, 24, 209–218.

Li, S., Sanna, A. & Anderson, J.M. (2011) Influence of temperature on pyrolysis of recycled organic matter from municipal solid waste using an activated olivine fluidized bed. *Fuel Processing Technology*, 92, 1776–1782.

Lin, Y.-H. & Yen, H.-Y. (2005) Fluidised bed pyrolysis of polypropylene over cracking catalysts for producing hydrocarbons. *Polymer Degradation and Stability*, 89, 101–108.

Lin, Y.-H. & Yang, M.-H. (2008) Tertiary recycling of polyethylene waste by fluidised-bed reactions in the presence of various cracking catalysts. *Journal of Analytical and Applied Pyrolysis*, 83, 101–109.

Liu, Y. & Liu, Y. (2005) Novel incineration technology integrated with drying, pyrolysis, gasification, and combustion of MSW and ashes vitrification. *Environmental Science & Technology*, 39, 3855–3863.

Lopez, U.A., de Marco, I., Caballero, B.M., Laresgoiti, M.F. & Adrados, A. (2010) Pyrolysis of municipal plastic wastes: Influence of raw material composition. *Waste Management*, 30, 620–627.

Lopez, U.A., de Marco, I., Caballero, B.M., Laresgoiti, M.F. & Adrados, A. (2011) Influence of time and temperature on pyrolysis of plastic wastes in a semi-batch reactor. *Chemical Engineering Journal*, 173, 62–71.

Lopez, U.A., de Marco, I., Caballero, B.M., Laresgoiti, M.F. & Adrados, A. (2012) Catalytic stepwise pyrolysis of packaging plastic waste. *Journal of Analytical and Applied Pyrolysis*, 96, 54–62.

Lucas, C., Szewczyka, D., Blasiaska, W. & Mochidab, S. (2004) High temperature air and steam gasification of densified fuels. *Biomass and Bioenergy*, 27 (6), 563–575.

Luo, S.Y., Xiao, B., Hu, Z.Q., Liu, S.M., Guan, Y.W. & Cai, L. (2010) Influence of particle size on pyrolysis and gasification of MSW in a fixed bed reactor. *Bioresource Technology*, 101, 6517–6520.

Lv, P., Yuan, Z.H., Ma, L.L., Wu, C.Z., Chen, Y. & Zhu, J.X. (2007) Hydrogen-rich gas production from biomass air and oxygen/steam gasification in a downdraft gasifier. *Renewable Energy*, 32 (13), 2173–2185.

Makishi, R. (September 17, 2014) *Material and Energy Recovery from Waste via Proven Gasification Technology*. Slide presentation, Energy from Waste Theatre.

Manos, G. (2006) Catalytic degradation of plastic waste to fuel over microporous materials. In: Scheirs, J. & Kaminsky, W. (eds.) *Feedstock Recycling and Pyrolysis of Waste Plastics*. New York, NY, Wiley. pp. 193–207.

Manyà, J.J., Sánchez, J.L., Ábrego, J., Gonzalo, A. & Arauzo, J. (2006) Influence of gas residence time and air ratio on the air gasification of dried sewage sludge in a bubbling fluidised bed. *Fuel*, 85, 2027–2033.

Mastral, F.J., Esparanza, E., Garcia, P. & Juste, M.J. (2002) Pyrolysis of high-density polyethylene in a fluidised bed reactor. Influence of the temperature and residence time. *Journal of Analytical and Applied Pyrolysis*, 63 (1), 1–15.

Mastral, J.F., Berrueco, C. & Ceamanos, J. (2006) Pyrolysis of high-density polyethylene in free-fall reactors in series. *Energy & Fuels*, 20 (4), 1365–1371.

Milne, T.A. & Evans, R.J. (1998) *Biomass Gasifier Tars: Their Nature, Formation, and Conversions*. Golden, CO, National Renewable Energy Laboratory. NREL/TP-570-25357.

Minutillo, M., Perna, A. & Bona, D.D. (2009) Modelling and performance analysis of an integrated plasma gasification combined cycle (IPGCC) power plant. *Energy Conversion and Management*, 50, 2837–2842.

Miranda, R., Pakdel, H., Roy, C. & Vasile, C. (2001) Vacuum pyrolysis of commingled plastics containing PVC ii. Product analysis. *Polymer Degradation and Stability*, 73, 47–67.

Miskolczi, N., Angyal, A., Bartha, L. & Valkai, I. (2009) Fuels by pyrolysis of waste plastics from agricultural and packaging sectors in a pilot scale reactor. *Fuel Processing Technology*, 90, 1032–1040.

Miskolczi, N., Borsodi, N., Buyong, F., Angyal, A. & Williams, P.T. (2011) Production of pyrolytic oils by catalytic pyrolysis of Malaysian refuse-derived fuels in continuously stirred batch reactor. *Fuel Processing Technology*, 92, 925–932.

Miskolczi, N., Ates, F. & Borsodi, N. (2013) Comparison of real waste (MSW and MPW) pyrolysis in batch reactor over different catalysts—Part II: Contaminants, char, pyrolysis oil properties. *Bioresource Technology*, 144, 370–379.

Miyoshi, F.J. (1998) *Resources & Environment*, 34 (14), 100–101.

Mohai, I. & Szépvölgyi, J. (2005) Treatment of particulate metallurgical wastes in thermal plasmas. *Chemical Engineering and Processing*, 44, 225–229.

Morris, M. (1998) *Electricity Production from Solid Waste Fuels Using advanced Gasification Technologies*. TPS Termiska Processer AB. Available from: www.tps.s.

Morris, M. & Waldheim, L. (1998) Energy recovery from solid waste fuels using advanced gasification technology. *Waste Management*, 18 (6–8), 557–564.

Mountouris, A., Voutsas, E. & Tassios, D. (2006) Solid waste plasma gasification: Equilibrium model development and exergy analysis. *Energy Conversion and Management*, 47, 1723–1737.

Moustakas, K. & Loizidou, M. (2010) Solid waste management through the application of thermal methods. In: Kumar, S. (ed.) *Waste Management*. InTech. ISBN: 978-953-7619-84-8. Available from: http://www.intechopen.com/books/waste-management/solid-waste-management-through-the-application-of-thermal-methods.

Moustakas, K., Fatta, D., Malamis, S., Haralambous, K. & Loizidou, M. (2005) Demonstration plasma gasification/vitrification systems for effective hazardous waste treatment. *Journal of Hazardous Materials*, 123, 120–126.

Muller-Hagedorn, M. & Bockhorn, H. (eds.) (2005) Selected papers presented at *Third International Symposium on Feedstock Recycling of Plastics & Other Innovative Plastic Recycling Techniques, Karlsruhe Institute of Technology, Karlsruhe, September 25–29, 2005*.

Murphy, A.B. (2003) Thermal plasma destruction of ozone-depleting substances: Technologies and chemical equilibrium, chemical kinetic and fluid dynamic modelling. *High Temperature Material Processes*, 7, 415–433.

Neeft, J.P.A., Knoef, H.A.M. & Onaji, P. (1999) *Behaviour of Tar in Biomass Gasification Systems. Tar Related Problems and Their Solutions*. November Report No. 9919, Energy from Waste and Biomass (EWAB), The Netherlands.

Nema, S.K. & Ganeshprasad, K.S. (2002) Plasma pyrolysis of medical waste. *Current Science India*, 83, 271–278.

Niessen, W.R. (2010) *Combustion and Incineration Processes: Applications in Environmental Engineering*. 4th edition. Boca Raton, FL, CRC Press, Taylor & Francis Group.

Nipathummakkul, N., Ahmed, I., Kerdsuwan, S. & Gupta, A.K. (2010) High-temperature steam gasification of waste water sludge. *Applied Energy*, 8 (12), 3729–3734.

NREL/TP-431-4988A.1 (1992) *Refuse-Derived Fuel, Data Summary of Municipal Solid Waste Management Alternatives; Volume I*. Golden, CO, SRI International for National Renewable Energy Laboratory. Available from: http://www.p2pays.org/ref/11/10516/refuse.html.

Nurul Islam, M., Alam Beg, M.R. & Rofiqul Islam, M. (2005) Pyrolytic oil from fixed bed pyrolysis of municipal solid waste and its characterization. *Renewable Energy*, 30, 413–420.

Okuwaki, A. (2004) Feedstock recycling of plastics in Japan. *Polymer Degradation and Stability*, 85, 981–988.

Okuwaki, A., Yoshioka, T., Asai, M., Tachibana, H., Wakai, K. & Tada, K. (2006) The liquefaction of plastic containers and packaging in Japan. In: Scheirs, J. & Kaminsky, W. (eds.) *Feedstock Recycling and Pyrolysis of Waste Plastics*. Mostoles, Spain, John Wiley & Sons.

Onwudili, J.A., Insura, N. & Williams, P.T. (2009) Composition of products from the pyrolysis of polyethylene and polystyrene in a closed batch reactor: Effects of temperature and residence time. *Journal of Analytical and Applied Pyrolysis*, 86, 293303.

Opinion Letter Regarding Plasma Arc gasification Options for management of Juneau's Waste SRS Engineers (2009) Available from: http://www.juneau.org/clerk/ASC/COTW/2009/documents/SCS_Engineers_Power_Point_Opinion_Letter_Regarding_Plasma_Arc_Gasification_and_Waste.pdf.

Oxtoby, D.W. (2002) *Principles of Modern Chemistry*. 5th edition. Thomson Brooks/Cole. ISBN: 0-03-035373-4.

Panda, A.K., Singh, R.K. & Mishra, D.K. (2009) Thermolysis of waste plastics to liquid fuel: A suitable method for plastic waste management and manufacture of value added products—A world prospective. *Renewable and Sustainable Energy Reviews*, 14 (1), 233–248.

Park, J.M., Kim, K.S., Hwang, T.H. & Hong, S.H. (2004) Three-dimensional modeling of arc root rotation by external magnetic field in non-transferred thermal plasma torches. *IEEE Transactions on Plasma Science*, 32, 479–487.

Petitpas, G., Rollier, J.D., Darmon, A., Gonzalez-Aguilar, J., Metkemeijer, R. & Fulcheri, L. (2007). A comparative study of non-thermal plasma assisted reforming technologies. *International Journal of Hydrogen Energy*, 32, 2848–2867.

Phan, A.N., Ryu, C., Sharifi, V.N. & Swithenbank, J. (2008) Characterisation of slow pyrolysis products from segregated wastes for energy production. *Journal of Analytical and Applied Pyrolysis*, 81 (1), 65–71.

Pinto, F., Costa, P., Gulyurtlu, I. & Cabrita, I. (1999a) Pyrolysis of plastic wastes: 1. Effect of plastic waste composition on product yield. *Journal of Analytical and Applied Pyrolysis*, 51, 39–55.

Pinto, F., Costa, P., Gulyurtlu, I. & Cabrita, I. (1999b) Pyrolysis of plastic waste: 2. Effect of catalyst on product yield. *Journal of Analytical and Applied Pyrolysis*, 51, 57–71.

Pinto, F., Franco, C., Andre, R.N., Miranda, M., Gulyurtlu, I. & Cabrita, I. (2002) Co-gasification study of biomass mixed with plastic wastes. *Fuel*, 81, 291–297.

Pinto, F., Franco, C., Andre, R.N., Tavares, C., Dias, M. & Gulyurtlu, I. (2003) Effect of experimental conditions on co-gasification of coal, biomass and plastic wastes with air/steam mixture in a fluidized bed system. *Fuel*, 82, 1967–1976.

PNNL (December 2008) *MSW to Liquid Fuels Synthesis, Volume 1: Availability of Feedstock and Technology*. Valkenberg, C., Gerber, M.A., Walton, C.W., Jones, S.B., Thomson, B.L. & Stevens, D.J., PNNL-18114. Prepared for US Department of Energy, by Pacific Northwest National Laboratory, Richland, WV.

Poiroux, R. & Rollin, M. (1996) High temperature treatment of waste: From laboratories to the industrial stage. *Pure and Applied Chemistry*, 68, 1035–1040.

Quaak, P., Knoef, H. & Stassen, H. (1999) *Energy from Biomass a Review of Combustion and Gasification Technologies*. Washington, DC, World Bank. xvii, 78.

4R Sustainability Inc. (April 2011) *Conversion Technology: A Complement to Plastic Recycling*. Technical Report.

Rajvanshi, A.K. (1986) Biomass gasification, Chapter 4. In: Yogi Goswami, D. (ed.) *Alternative Energy in Agriculture*. Vol. II. Boca Raton, FL, CRC Press. pp. 83–102.

Reed, T.B. & Gaur, S. (2001) *A Survey of Biomass Gasification*. ISBN: 1-890607-13-4180.

Rezaiyan, J. & Cheremisinoff, N.P. (2005) *Gasification Technologies: A Primer for Engineers and Scientists*. Boca Raton, FL, CRC Press. p. 336.

Ricketts, B., Hotchkiss, R., Livingston, B. & Hall, M. (2002) Technology status review of waste/biomass cogasification with coal. In: *IChemE Fifth European Gasification Conference, 8–10 April 2002, Noordwijk, The Netherlands*.

Roos, C. (2010) *Clean Heat and Power Using Biomass Gasification for Industrial and Agricultural Projects*. U.S. Department of Energy. pp. 1–9.

RTI International (January 2012) *Environmental and Economic Analysis of Emerging Plastic Conversion Technologies, Final Project Report*. Prepared for American Chemical Council, RTI project No. 0212876.000.

Rutberg, G., Bratsev, A.N., Safronov, A.A., Surov, A.V. & Schegolev, V.V. (2002) The technology and execution of plasma chemical disinfection of hazardous medical waste. *IEEE Transactions on Plasma Science*, 30, 1445–1448.

Rutberg, Ph.G., Bratsev, A.N., Kuznetsov, V.A., Popov, V.E., Ufimtsev, A.A. & Shtengel', S.V. (2011) On efficiency of plasma gasification of wood residues. *Biomass and Bioenergy*, 35, 495–504.

Rutkowski, P. (2012) Chemical composition of bio-oil produced by co-pyrolysis of biopolymer/polypropylene mixtures with K_2CO_3 and $ZnCl_2$ addition. *Journal of Analytical and Applied Pyrolysis*, 95, 38–47.

Saffarzadeh, A., Shimaoka, T., Motomura, Y. & Watanabe, K. (2009a) Characterization study of heavy metal-bearing phases in MSW slag. *Journal of Hazardous Materials*, 164, 829–834.

Saffarzadeh, A., Shimaoka, T., Motomura, Y. & Watanabe, K. (2009b) Petrogenic characteristics of molten slag from the pyrolysis/melting of MSW. *Waste Management*, 29, 1103–1113.

Sahraei-Nezhad, F. & Akhlagi-Boozani, S. (2010) *Production and Gasification of Waste Pellets*. Thesis submitted to School of Engineering. Boras, University of Boras.

Sandquist, J. (2011) *Biomass Gasification in Norway*. SINTEF Energy. Available from: www.ieatask33.org/app/webroot/files/file/2011/Norway.pdf.

Saravanakumar, A., Haridasan, T.M., Reed, T.B. & Bai, R.K. (2007) Experimental investigations of long stick wood gasification in a bottom lit updraft fixed bed gasifier. *Fuel Processing Technology*, 88 (6), 617–622.

Sawaguchi, T., Inami, T., Kuroki, T. & Ikemura, T. (1980) Studies on thermal degradation of synthetic polymers. 12. Kinetic approach to intensity function concerning pyrolysis condition for polyethylene low polymer. *Industrial & Engineering Chemistry Process Design and Development*, 19 (1), 174–179.

Sawaguchi, T., Inami, T., Kuroki, T. & Ikemura, T. (1981) Studies on thermal degradation of synthetic polymers. XV. Estimation of the product yield on the basis of intensity function for thermal gasification of isotactic and atactic polypropylenes. *Journal of Applied Polymer Science*, 26 (4), 1267–1274.

Scheirs, J. (2006) Overview of commercial pyrolysis processes for waste plastics. In: Scheirs, J. & Kaminsky, W. (eds.) *Feedstock Recycling and Pyrolysis of Waste Plastics*. New York, NY, Wiley. pp. 381–433.

Scheirs, J. & Kaminsky, W. (eds.) (2006) *Feedstock Recycling and Pyrolysis of Waste Plastics*. New York, NY, Wiley.

Scott, D.S., Czemik, S.R., Piskorz, J. & Radlein, D.S.A.G. (1990) Fast pyrolysis of plastic wastes. *Energy & Fuels*, 4 (4), 407–411.

Scottish Agricultural Website (2002) Available from: http://www.sac.ac.uk/envsci/External/WillowPower/Conversn.htm.

Seifert, H. & Vehlow, J. (2009) Some observations on the operation of waste treatment technologies in Japan at the Task 36 meeting in Fukuoka, Japan in November 2009; Appendix 2: Study Tour of Japan; Karlsruhe Institute of Technology, Institute for Technical Chemistry/Div. of Thermal Waste Treatment.

Shah, J., Rasul Jan, M., Mabood, F. & Jabeen, F. (2010) Catalytic pyrolysis of LDPE leads to valuable resource recovery and reduction. *Energy Conversion and Management*, 51, 2791–2801.

Smolders, K. & Baeyens, J. (2004) Thermal degradation of PMMA in fluidised beds. *Waste Management*, 24, 849–857.

Soheilian, R., Davies, A., Anaraki, S.T., Zhuo, C. & Levendis, Y.A. (2013) Pyrolytic gasification of post-consumer polyolefins to allow for 'clean' premixed combustion. *Energy & Fuels*, 27, 4859–4868. Reprinted with permission from ACS, © 2013 American Chemical Society.

Solenagroup. Available from: http://www.solenagroup.com.

Solid Waste Association of North America (SWANA) Applied Research Foundation (2011) *Waste Conversion Technologies*.

Sorum, L., Gronli, M.G. & Hustad, J.E. (2001) Pyrolysis characteristics and kinetics of municipal solid wastes. *Fuel*, 80 (9), 1217–1227.

Slapak, M.J.P., Kasteren, J.M.N.V. & Drinkenburg, A.A.H. (2000) Design of a process for steam gasification of PVC waste. *Resources, Conservation and Recycling*, 30, 81–93.

Spliethoff, H. (2010) *Power Generation from Solid Fuels*. Berlin – Heidelberg, Springer.

Stein, W. & Tobiasen, L. (March 2004) *Review of Small-Scale Waste to Energy Conversion Systems, IEA Bioenergy Agreement—Task 36, Work Topic 4*. Available from: www.researchgate.net/publications.PublicPostFileLoader.ht.

Steiner, C., Kameda, O., Oshita, T. & Sato, T. (2002) EBARA's fluidized bed gasification: Atmospheric 2 × 225 t/d for shredding residues recycling and two-stage pressurized 30 t/d for ammonia synthesis from waste plastics. In: *Proceedings of Second International Symposium on Feedstock Recycle of Plastics and Other Innovative Plastics Recycling Techniques, Ostend, 8–11 September 2002*.

Stelmachowski, M. (2010) Thermal conversion of waste polyolefins to the mixture of hydrocarbons in the reactor with molten metal bed. *Energy Conversion and Management*, 51 (10), 2016–2024.

Surisetty, V.R., Kozinski, J. & Dalai, A.K. (2012) Biomass, availability in Canada, and gasification: An overview. *Biomass Conversion and Biorefinery*, 2, 73–85.

Surma, J. (2012) *Conversion of Municipal Solid Waste into Clean Energy Products Using the InEnTec Plasma Enhanced Melter®* . Available from: http://www.epa.gov/region9/organics/symposium/2012/2012-cba-jeff-surma.pdf.

Syamsiro, M., WuHu, S., Komoto, S., Cheng, P., Noviasri, P., Prawisudha, K. & Yoshikawa, K. (2013) Co-production of liquid and gaseous fuels from PE and PS in a continuous sequential pyrolysis and catalytic reforming system. *Energy and Environment Research*, 3 (2), 90–106.

Syamsiro, M., Cheng, S., Hu, W., Saptoadi, H., Pratama, N.N., Trisunaryanti, W. & Yoshikawa, K. (2014) Liquid and gaseous fuels from waste plastics by sequential pyrolysis and catalytic reforming processes over Indonesian natural zeolite catalysts. *Waste Technology*, 2 (2), 44–51.

Tang, C., Wang, Y.-Z., Zhou, Q. & Zheng, L. (2003) Catalytic effect of Al–Zn composite catalyst on the degradation of PVC-containing polymer mixtures into pyrolysis oil. *Polymer Degradation and Stability*, 81, 89–94.

Themelis, N.J. & Mussche, C. (July 9, 2014) *2014 Energy and Economic Value of Municipal Solid Waste, Including Non-Recycled Plastics (NRP), Currently Landfilled in the Fifty States*. New York, NY, Earth Engineering Center, Columbia University.

Thomas, M. (2004) Novel and Innovative pyrolysis and gasification technologies for energy efficient and environmentally sound MSW disposal. *Waste Management*, 24 (1), 53–79.

Tiikma, L., Johannes, I. & Luik, H. (2006) Fixation of chlorine evolved in pyrolysis of PVC waste by Estonian oil shales. *Journal of Analytical and Applied Pyrolysis*, 75, 205–210.

Tillman, D.A. (1991) *The Combustion of Solid Fuels and Wastes*. San Diego, CA, Academic Press.

Tukker, A., de Groot, H., Simons, L. & Wiegersma, S. (1999) *Chemical Recycling of Plastic Waste: PVC and Other Resins*. European Commission, DG III, Final Report, STB-99-55 Final, Delft, The Netherlands.

TwE (August 2014) *Gasification Technologies Review—Technology, Resources and Implementation Scenarios*. Final Revised Report. Prepared by Talent with Energy for the City of Sydney's Advanced Waste Treatment Master Plan.

Tzeng, C.-C., Kuo, Y.-Y., Huang, T.-F., Lin, D.-L. & Yu, Y.-J. (1998) Treatment of radioactive wastes by plasma incineration and vitrification for final disposal. *Journal of Hazardous Materials*, 58, 207–220.

U.C. Riverside (June 21, 2009) *Evaluation of Emissions from Thermal Conversion Technologies Processing Municipal Solid Waste and Biomass*. Prepared for BioEnergy Producers Association, Los Angeles, CA, by University of California, Riverside, CA. pp. 38–41.

UNEP (2009) *Converting Waste Plastics into a Resource—Compendium of Technologies*. Osaka/Shiga, Japan, Intl. Environment Technology Center.

URS Corporation (2005) *Conversion Technology Evaluation Report*. Tech. Rep., The County of Los Angeles Department of Public Works and The Los Angeles County Solid Waste Management Committee/Integrated Waste Management Task Force's Alternative Technology Advisory Subcommittee, Los Angeles, CA.

USDOE Report (2002) *Gasification Markets and Technologies—Present and Future: An Industry Perspective—DOE/FE-0447.*

Van Wylen, G.J., Sonntag, R.E. & Borgnakke, C. (1994) *Fundamentals of Classical Thermodynamics*. 4th edition. New York, NY, Wiley. xii, 852 pp.

Velghe, I., Carleer, R., Yperman, J. & Schreurs, S. (2011) Study of the pyrolysis of municipal solid waste for the production of valuable products. *Journal of Analytical and Applied Pyrolysis*, 92, 366–375.

Wang, Y. & Kinoshita, C.M. (1993) Kinetic model of biomass gasification. *Solar Energy*, 51 (1), 19–25.

Watanabe, T. & Shimbara, S. (2003) Halogenated hydrocarbon decomposition by steam thermal plasmas. *High Temperature Material Processes*, 7, 455–474.

Watanabe, T., Tsuru, T. & Takeuchi, A. (2005) Water plasma generation under atmospheric pressure for effective CFC destruction. In: Mostaghimi, J., et al. (eds.) *Proc. 17th Intl. Symp. Plasma Chemistry, Toronto, Canada*. University of Toronto.

Weissman, R. (1997) Recycling of mixed plastic waste by the Texaco gasification process. In: Hoyle, W. & Karsa, D.R. (eds.) *Chemical Aspects of Plastics Recycling*. Cambridge, The Royal Society of Chemistry Information Services.

Welendziewski, J. (2002) Engine fuels derived from waste plastics by thermal treatment. *Fuel*, 81, 473–481.

Westerhout, R.W.J., Waanders, J., Kuipers, J.A.M. & van Swaaij, W.P.M. (1998) Recycling of polyethene and polypropene in a novel bench-scale rotating cone reactor by high-temperature pyrolysis. *Industrial & Engineering Chemistry Research*, 37 (6), 2293–2300.

WGT (2002) *Waste Gas Technology Energy from Waste*. Available from: http://www.wgtuk.com/ukindex.html.

Williams, E.A. & Williams, P.T. (1997a) Analysis of products derived from the fast pyrolysis of plastic waste. *Journal of Analytical and Applied Pyrolysis*, 40–41, 347–363 (Proc. of the 12th International Symposium on Analytical and Applied Pyrolysis, Oct 14–18 1996).

Williams, E.A. & Williams, P.T. (1997b) Pyrolysis of individual plastics and a plastic mixture in a fixed bed reactor. *Journal of Chemical Technology and Biotechnology*, 70 (1), 9–20.

Williams, P.T. (April 2003) *Recycling Scrap Tyres to Valuable Products*. Green Chemistry. G20–G23.

Williams, P.T. (2006) Yield and composition of gases and oils/waxes from the feedstock recycling of waste plastic. In: Scheirs, J. & Kaminsky, W. (eds.) *Feedstock Recycling and Pyrolysis of Waste Plastics*. Leeds, John Wiley & Sons, Ltd. pp. 286–313.

Williams, P.T. (2012) *Fuels, Chemicals and Materials from Waste*. 2012 ECG Distinguished Guest Lecture, The Royal Society of Chemistry: Environmental Chemistry Group, Chemistry Centre, Burlington House, London, 14th March 2012, in ECG Bulletin, Royal Society of Chemistry, July 2012, p. 5.

Williams, P.T. & Williams, E.A. (1998) Recycling plastic waste by pyrolysis. *Journal of the Institute of Energy*, 71 (487), 81–93.

Williams, P.T. & Williams, E.A. (1999) Interaction of plastics in mixed-plastics pyrolysis. *Energy & Fuels*, 13 (1), 188–196.

Williams, P.T. & Slaney, E. (2007) Analysis of products from the pyrolysis and liquefaction of single plastics and waste plastic mixtures. *Resources, Conservation and Recycling*, 51 (4), 754–769.

Wilson, B.W. & Wilson, B.R. (October 2013) *A Comparative Assessment of Commercial Technologies for Conversion of Solid Waste to Energy*. Enviro Power Renewable, Inc. Available from: http://www.itigroup.co/uploads/files/59_comparative_WTE-Technologies-Mar_2014.pdf.

Wolf, C.J., Heanley, C.P. & Cashell, P.V. (1998) Plasma arc system for vitrification of LLRW. In: *Proc. 19th US Department of Energy Low-Level Radioactive Waste Management Conf. Salt Lake City, UT, USA*.

Wong, S.L., Ngadi, N., Abdullah, T.A.T. & Inuwa, I.M. (2015) Current state and future prospects of plastic waste as a source of fuel: A review. *Renewable and Sustainable Energy Reviews*, 50, 1167–1180.

WSP (2013) *Review of State-of-the-Art Waste-to-Energy Technologies, Stage 2—Case Studies*. Prepared by Kevin Whiting, WSP Environmental Ltd, WSP House, London WC2A 1AF, UK.

Wu, J., Chen, T., Luo, X., Han, D., Wang, Z. & Wu, J. (2014) TG/FTIR analysis on co-pyrolysis behavior of PE, PVC and PS. *Waste Management*, 34, 676–682.

Xiao, G., Ni, M.J., Chi, Y., Jin, B.S., Xiao, R., Zhong, Z.P. & Huang, Y.J. (2009) Gasification characteristics of MSW and an ANN prediction model. *Waste Management*, 29, 240–244.

Yamada, S., Shimizu, M. & Miyoshi, F. (July 2004) *Thermoselect Waste Gasification and Reforming Process*. JFE Technical Report, No. 3.

Yan, G., Jing, X., Wen, H. & Xiang, S. (2015) Thermal cracking of virgin and waste plastics of PP and LDPE in a semibatch reactor under atmospheric pressure. *Energy Fuels*, 29 (4), 2289–2298.

Yang, H., Yan, R., Chen, H., Lee, D.H., Liang, D.T. & Zheng, C. (2006) Pyrolysis of palm oil wastes for enhanced production of hydrogen rich gases. *Fuel Processing Technology*, 87, 935–942.

Yassin, L., Lettieri, P., Simons, S. & Germanà, A. (2009) Techno-economic performance of energy-from-waste fluidized bed combustion and gasification processes in the UK context. *Chemical Engineering Journal*, 146, 315–327.

Yoshikawa, K. (2004) Commercial demonstration of a small-scale gasification and power generation from MSW. In: *Proc. 2nd Intl. Energy Conversion Engineering Conference & Exhibit, Providence, RI, 16–19 August 2004*. Reston, VA, American Institute of Aeronautics and Astronautics. AIAA No. 2004-5649, ISSN: 10877215.

Yoshikawa, K. (2010) *Gasification and Liquifaction—Alternatives to Incineration for MSW Recycling Developed and Commercialized in Japan*. Available from: www.seas.columbia.edu/earth/wtert/meet2010/procedings/presentations/YOSHIKAWA.pdf.

Yoshikawa Lab. Website: *Pyrolytic Oil from Waste Plastics*. Yoshikawa Laboratory, Tokyo Institute of Technology. Available from: http://yoshikawalab.org/index.php?option=com_content&view=article&id=121&Itemid=200&lang=en.

Yoshioka, T., Gause, G., Eger, C., Kaminsky, W. & Okuwaki, A. (2004) Pyrolysis of polyethylene terephthalate in a fluidised bed plant. *Polymer Degradation and Stability*, 86, 499–504.

Young, G.C. (2010) *Municipal Solid Waste to Energy Conversion Process.* New York, NY, John Wiley & Sons, Inc.

Young, N.C., Seong, C.K. & Kunic, Y. (2011) Pyrolysis gasification of dried sewage sludge in a combined screw and rotary kiln gasifier. *Applied Energy*, 88 (4), 1105–1112.

Zhang, L., Xu, C. & Champagne, P. (2010) Overview of recent advances in thermo-chemical conversion of biomass. *Energy Conversion and Management*, 51, 969–982.

Zhao, W., Hasegawa, S., Fujita, J., Yoshii, F., Sasaki, T., Makuuchi, K., et al. (1996) Effects of zeolites on the pyrolysis polypropylene. *Polymer Degradation and Stability*, 53, 129–135.

Zhao, Y., Sun, S., Zhou, H., Sun, R., Tian, H., Luan, J. & Qian, J. (2010) Experimental study on sawdust air gasification in an entrained-flow reactor. *Fuel Processing Technology*, 91, 910–914.

Zhao, L., Chen, D.Z., Wang, J.H., Ma, X.B. & Zhow, G. (2011) Pyrolysis of waste plastics and whole combustible components separated from MSW: Comparison of products and emissions. In: *Proc. 13th Intl. Waste Management and landfill Symp. 3–6 Oct. 2011, Sardinia.* pp. 117–118.

Zhou, C. (2014) *Gasification and Pyrolysis: Characterization and Heat transfer Phenomena During Thermal Conversion of Municipal Solid Waste.* Doctoral Dissertation. Stockholm, Sweden, Division of Energy and Furnace Technology, Department of Materials Science and Engineering, School of Industrial Engineering & Management, KTH-Royal Institute of Technology.

Zhou, C., Zhang, Q., Arnold, L., Yang, W. & Blasiak, W. (2013): A study of the pyrolysis behaviours of pelletized recovered municipal solid waste fuels. *Applied Energy*, 107, 173–182.

Zhuo, C. (2009) *Synthesis of Carbon Nanotubes from Waste Polyethylene Plastics.* Available from: http://hdl.handle.net/2047/d20000794.

Zhuo, C., Hall, B., Richter, H. & Levendis, Y. (2010) Synthesis of carbon nanotubes by sequential pyrolysis and combustion of polyethylene. *Carbon*, 48 (14), 4024–4034.

Websites: These Company websites describe technologies which, after initial trials and testing, have been stalled for reasons unknown.

www.agilyx.com

www.climaxglobalenergy.com

www.respolyflow.com

http://Cynarplc.com

http://www.vadxx.com/

www.plastics2oil.com

Chapter 5

MSW thermal conversion plants: Case Studies

5.1 AEB, AMSTERDAM, THE NETHERLANDS

The moving grate combustion plant at Amsterdam is the world's biggest one, comprises six incineration lines and processes the household, industrial and commercial waste from Amsterdam and surroundings amounting to 1.5 million tonnes each year (de Waart 2009).

The original Afval Energie Bedrijf (AEB) plant was built and operated in Amsterdam in 1917 with four combustion lines supplied by Widmer & Enrst (W&E). The plant was refurbished and the air pollution control systems upgraded by Lurgi by installing an extra wet scrubbing complying the latest emission policies which set the standard for the current European emission targets in 1993. Two new lines (MARTIN horizontal grates) were added in 2007 and the performance of the steam cycle for the new two lines is optimised by including a reheat cycle. The two new incineration lines of the plant were designed for maximum recovery each with a thermal capacity of 93.3 MW. At an average calorific value of 10 MJ/kg, this corresponds to a nominal throughput of 33.6 tonnes/hr of waste, or a total of 1,600 t/day, fairly large. The large size has the major advantage that fluctuations in the waste processed can be easily absorbed. Moreover, variations in the composition of the waste have less effect and the boilers can operate steadily and without disruptions.

The two boilers are designed to produce a thermal output of 93.6 MW and steam at 130 bar and up to 460°C (load point B) and 480°C (load point C) with a thermal efficiency of 87.14%. One steam turbine and generator is shared between the two lines to take the total steam flow and produce 66 MWe. The electrical conversion efficiency during the acceptance test was 30.6%.

The plant is amongst the world's cleanest (emissions <20% EPA limits), with an overall solids recycling rate of 95%, and with no waste water discharge. The plant maximises recovery of materials for reuse, such as bottom ash and fly ash, as well as producing calcium chloride and gypsum as secondary by-products of the flue gas cleaning process. The annual availability is reported to be greater than 90%. The electricity produced is pumped to the grid and the heat energy exported to the local district heating network.

The crane system lifts the waste from the bunker located in a closed building and transports to the feed chute, and the feed rams in turn push the waste onto the combustion grate in amounts determined by the combustion control system. Each

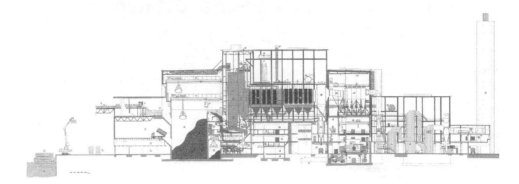

Figure 5.1 Schematic of Moving grate process incinerator (High resolution image can be consulted online at https://www.crcpress.com/Energy-Recovery-from-Municipal-Solid-Waste-by-Thermal-Conversion-Technologies/Reddy/9781138029552) (source: WSP 2013).

feed ram is driven by a hydraulic cylinder. The combustion control system prescribes the cycle time, stroke length and stroke speed to achieve uniform combustion on the grate.

The grates (four W&E and two MARTIN horizontal grates) in the six incineration lines consist of alternating rows of fixed and moving grate bars; and as the neighboring moving grate bar rows make a counter movement, the waste is effectively transported and mixed to ensure good burnout. The horizontal grate is of modular design. The length of each module is fixed but the width may vary depending on the design. Each module has its own drive and supply of underfire air, both controlled separately. A typical grate configuration consists of 3 modules in the direction of waste flow. The waste is burned on horizontal grates that are identical. The grate is of the double-motion over-thrust type and comprises three parallel runs, each with seven air zones. The grate bar design including grate cooling system and air injection is explained in the WSP report (2013).

The bottom ash produced during waste incineration drops down the slag shaft into a water-filled de-slagger. Each grate has two of these (one per grate run). Closed oscillating conveyors transport this to the slag bunker. A tertiary air exhaust system sucks water vapour out of the de-slagging system. The air injection has been designed using CFD in order to assure the optimum turbulence in the combustion zone and a stable and smooth flue gas flow in the boiler. Secondary air is the recirculated flue gas drawn from the bag filter exit for efficiency and NO_x reduction.

The high efficiency boiler has three empty radiation chambers and a horizontal convection section with tube banks for the steam superheaters and economisers. To realize the maximum reliability and capacity, the furnace is designed spacious with large heat exchange surfaces. The designed velocity of the flue gas and its temperature are low before it reaches the first steam superheater banks. The greater volume helps reduce the amount of dust in the flue gas in addition to lowering the temperature. It also lowers the corrosion effects in the superheater tubes and increases the reliability of the steam superheater banks and the economisers. The extra capacity also contributes to the flexibility needed to separately generate saturated steam for the intermediary

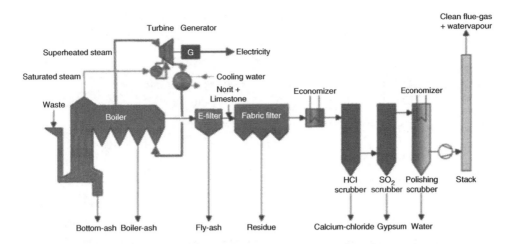

Figure 5.2 Schematic of the block flow diagram of the entire Plant (de wart 2009).

superheating. The first two boiler radiation chambers are fitted with Inconel cladding; so also the membrane walls under the brickwall. This is because the temperature of the membrane walls reaches around 340°C instead of the typical 270°C due to high pressure.

The boiler has been designed allowing an empty space between the superheater and economiser sections to accommodate in future an extra tube bank. This facilitates to achieve an even higher steam temperature of 480°C which can further enhance the plant's *total efficiency*. It is also possible to connect the superheater tube banks entirely for counterflow. This addition to obtain higher steam temperature of 480°C, however, depends on whether the degree of wear through corrosion/erosion remains within practical limits. The block flow diagram of the entire plant is shown in Figure 5.2.

To increase the overall efficiency, the following steps were taken:

(a) the steam pressure has been raised from 40 to 125–130 bar, made possible by reheating of the steam from the high-pressure stage of the turbine using saturated steam from the steam drum of the boiler instead of usual flue gases (Fig. 5.2);

(b) minimizing the oxygen level in the combustion zone from the typical 8–11% to 6% by flue gas recirculation, also resulting in reduced volume of flue gas by about 40% and minimizing the amount of heat lost via the stack;

(c) maximizing the use of flue gas energy by (i) lowering the flue gas temperature exiting the boiler to 180°C instead the normal 200–240°C by enlarging the first stage economiser; (ii) fixing a corrosion-resistant heat exchanger before the first flue gas scrubber (the quench) to preheat the condensate; and (iii) cooling the flue gases in the last polishing scrubber until the water condenses (This heat is also used for preheating the condensate); and

(d) minimizing the steam pressure after the turbine. (Reheating the turbine steam and cooling the condenser with sea water facilitate to have a very low steam pressure after the turbine, which enables maximum turbine efficiency).

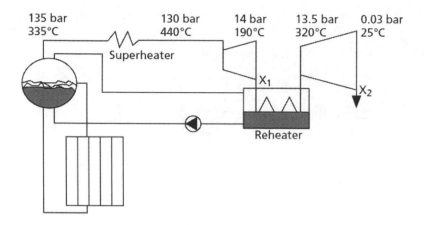

Figure 5.3 Steam reheating outline (source: Redrawn from de Waart 2009).

Flue gas cleaning and recovery of materials: The schematic of the system along with product recovery is also shown in Figure 5.3. SNCR is used for reducing NO_x; and to optimise efficiency compressed air instead of steam is used for injecting ammonia, diluted with softened water. For the pre-separation of fly ash, an ESP has been fitted after the boiler outlet which in combination with the boiler's two-way ash conveyance system allows the maximum amount of fly ash to be reused. This was done for economical optimization.

A fabric filter into which fine powdered activated carbon or blast furnace coke is blown in as the adsorption medium is fitted to remove fine particles. Limestone powder is added to the coke injection to eliminate the risk of fire and explosions. It also forms a filtering layer on the filter bags. The coke adsorption medium separates out dioxins and furans at the beginning of the flue gas cleaning process. Heavy metal content is also reduced to such an extent that the products can be reused.

The HCl and SO_2 scrubbers are the next stage to clean acid gases and ammonia from the flue gas. The spray from the HCl scrubber is sent to the quench as a concentrated HCl solution. Lime-milk solution is added as a neutralizer into the SO_2 scrubber, which at pH 6, reacts with SO_2 in the scrubber to form gypsum slurry. By using a fabric filter, relatively clean gypsum is produced. A centrifuge separates the gypsum from the slurry as a dry product that is stored for reuse.

Next follows a separate polishing scrubber consisting of a packed bed over which water circulates that is cooled by a water-water heat exchanger. This cooling of the flue gas leads to condensation of the water in the flue gas, further reducing emissions. The recovered heat is used in the first stage for preheating the condensate. By super-cooling the flue gases, the polishing scrubber with ECO3 produces virtually pure condensation water that can be reused in the flue gas cleaning system, mainly in the HCl scrubber and the quench.

Finally, the purified flue gas is kept at under pressure by an induced draught fan. This runs 'wet' in the saturated flue gases that pass through a drip tray and emissions monitoring equipment to the stack.

Figure 5.4 Photo of the Amsterdam Waste Fired Power Plant (de Waart 2009).

Residues treatment and usage: The system is so designed that bottom ash, boiler ash and fly ash can be separated and processed for beneficial uses. For example, bottom ash is used in sand-lime bricks, and concrete; and fly ash in asphalt concrete.

Similarly, HCl is processed into calcium chloride and SO_2 into gypsum for reuse. The residues from the fabric filter (hazardous waste) only have to be disposed of in a hazardous landfill. The blowdown from the quench and hydrochloric acid scrubber is neutralised with limestone. The pH is increased further by the addition of calcium hydroxide and caustic soda.

The addition of hydrogen sulfide produces a precipitation of heavy metals as sulfides. This slurry is separated and transported to the filter press. An evaporation plant strips any ammonia from the solution and recycles it for ammonia injection in the boiler (SNCR). A sand filter, ion exchanger and active carbon filter together constitute the polishing stage that guarantees the quality of the salt, which can be marketed as a product.

All water produced is internally recycled for use in other process steps or evaporated. As such, the plant produces no waste water. As the evaporation is low, a clean brine is discharged. Studies to utilize this brine in the concrete industry as an economical proposition are currently in progress.

*In summary, the newest two lines of the Amsterdam moving grate combustion plant really are **state-of-the-art and best example of available Incineration technology.** Not only does the process produce electricity with a net efficiency of >30%, the highest of any WtE combustion plant in the world, but the plant also maximises recovery of materials for re-use such as bottom ash and fly ash, as well as producing calcium chloride and gypsum as secondary by-products of the flue gas cleaning process. The annual availability is reported to be >90%.* (References: de Waart 2009, WSP 2013, Sietse Agema 2016)

5.2 ISSEANE POWER PLANT, PARIS, FRANCE

The Waste to Energy plant (known as Isseane) owned by Syctom Paris and commissioned in 2007 is located at a few kms from the centre of Paris. The state-of-the-art design of the plant, the largest in France processing 460,000 TPA of MSW with high efficiency, involved high capital costs compared to a traditional one, and provides electricity and heat. The plant is designed to treat waste of calorific value ranging from 8 to 12 MJ/kg.

It is a two-line water-cooled grate incinerator, Hitachi Zosen Inova design, incorporating a conventional steam cycle. Waste is combusted on one of the grates. The size of each grate is 10 m × 10 m and has five zones, the initial three being water cooled by Aquaroll. Hot gases exit the furnace into a secondary combustion chamber and are transferred through a 4-pass horizontal boiler. The sophisticated combustion control system includes an infrared camera to monitor combustion and enable adjustments to be made to respond to varying waste feedstock and optimize burn-out.

The flue gases passing through the boiler gets cooled from 1100°C to around 200°C and raises steam. The two lines produce a total of 200 tonnes per hour of superheated steam at 400°C and 50 bar which is passed to a single Alstom 56 MW condensing turbine. The turbine incorporates an off take for low pressure steam, most of it used to generate hot water for supply via a district heating network. A water cooled condenser uses water from the River Seine where the plant is located, to condense the turbine exhaust gases, which is subsequently returned to the river. The Plant's lay-out is shown schematically in Figure 5.5a.

Flue gas cleaning system: The plant is fitted with state-of-the-art pollution control system to meet the EU directive for the emissions limits. Flue gases are first passed through an ESP which removes around 99% of particulate matter. A dry sorption sodium bicarbonate system neutralises acid gases, in particular SO_2, and an activated carbon system further removes pollutants and dioxins. A fabric filter removes particulate matter not captured by the ESP and products from the acid gas removal and activated carbon systems. Finally, a low temperature SCR deNOx system removes most of NOx in the flue gases by injection of ammonia in the presence of a catalyst at around 220°C. The cleaned gas then escapes via the two low profile stacks. In order to prevent a plume from forming, a reheat gas burner is included to raise the temperature of the exit cleaned flue gas (maximum of 290°C) to prevent the formation of water vapour at the stack exit, and gases are emitted at a temperature not less than 200°C and a speed of around 30 m/sec to ensure adequate dispersion.

The combination of ESP and fabric filter is unusual as most plants have one or the other; at this plant, both are employed in order to meet the very strict standards relating to particulate emissions, which are more than three times more stringent than Waste incineration Directive of EU (WID). Cost-wise, SNCR systems are usually preferred, though SCR is more effective and required at the plant to meet the stringent emission limit for NOx.

Residues processing: Bottom ash from the grate is collected and transported by boat along the river Siene to treatment centres for processing. Fly ash from the ESP and fabric filter is collected in silos and treated prior to disposal in a hazardous landfill. The fabric filter residue is high in sodium compounds, and sodium carbonate is recovered where possible. The remaining residues are then solidified by mixing with cement

Issy-Les-Moulineaux / France

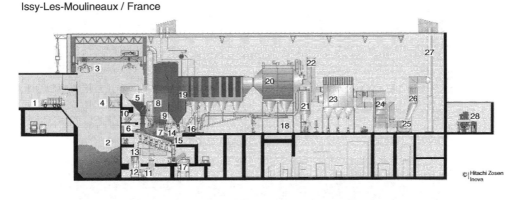

Waste receiving and storage	Combustion and boiler		Flue gas treatment	Energy recovery
1 Tippin hall	5 Feed hopper	12 Primary air preheater	20 Electrostatic precipitator	28 Turbine and generator
2 Waste pit	6 Ram feeder	13 Primary air distribution	21 Sodium bicarbonate silo	
3 Waste crane	7 Hitachi Zosen Inova grate	14 Secondary air fan	22 Flue gas entrainment duct	
4 Loader control cabin	8 Infrared camera	15 Secondary air/recirculated flue gas injection	23 Fabric filter	
	9 Start-up and support burners	16 Recirculation fan	24 SCR catalyst	
	10 Burner fan	17 Bottom ash transport	25 Induced-draft fan	
	11 Primary air fan	18 Ash conveyor	26 Silencer	
		19 Four-pass boiler	27 Stack	

Figure 5.5a Isseane WtE Power plant layout (reproduced with the permission of Hitachi Zosen Inova AG; Dr. Michael Keunecke).

and special additives that help stabilize the pollutants. The bottom ash and metals produced in 2011 are: (a) 85,583 tons of bottom ash; (b) 7,226 tons of ferrous metals; and (c) 726 tons of nonferrous metals. The bottom ash and metals are recycled for reuse, and fly ash is treated before disposal in a hazardous landfill.

Power and Heat production: The turbine generator has a rated output of 52 MW which can only be achievable when no steam is extracted for the district heating network. The performance data for 2011: 459,772 tonnes of waste throughput; 132,336 MWh of electricity; 87,684 MWh of electricity exported; and 513,331 MWh of steam (WSP 2013). Although accurate efficiency is not assessed as no data on the actual waste calorific value is available, it is implied that an annual average gross electrical efficiency is around 10% and a thermal efficiency of around 40%. Despite the low electrical output the overall efficiency is around 50%, which is much higher compared to a plant exporting only electricity.

Continuous monitoring of air emissions during operation of the plant has established that they are well below the stipulated limits. The process water effluent is discharged to sewer following treatment in an internal effluent treatment centre to meet discharge limits; and there is no waste water discharge to the River Seine.

Unique Building design: Widely known for its innovative design, the plant was planned to minimize local environmental and aesthetic impacts. In order to unify with the surrounding building structures, the plant is buried 31 meters below the ground

Figure 5.5b Photo of the Isseane Plant (reproduced with the permission of Hitachi Zosen Inova AG –
photo sent by Dr. Michael Keunecke).

and stands only 21 meters above ground level (approximately equivalent to a 6-story building), and the chimney has been designed to protrude only 5 m so that the stack is not visible on the Paris skyline. The plant building more closely resembles an office construction. The plant is clad in masonry and has a green roof and the heavy tree planting that helps to mask the building as well as encourages biodiversity. There is no visual plume from the stacks due to the flue gas reheat system (Photo).

> *In summary, the ISSEANE plant is a state-of-the-art facility and a major feat of engineering. The plant is sunk about 31 metres into the bank of the River Seine with all the associated hydro- geological challenges of building the plant there. The exhaust gas chimneys project only 5metres above the building, but in order to do this the plant had to guarantee emission limits to air of 50% of the WID values for all pollutants.* (Reference: Hitachi Zosen Inova website: Plants in Operation; WSP 2013)

5.3   ZABALGARBI PLANT (SENER-2), BILBAO, SPAIN

The Zabalgarbi waste to energy plant near Bilbao is a conventional grate incinerator with an advanced energy recovery system. The steam produced by the waste heat recovery boiler of the Plant is used within a Combined Cycle Gas Turbine (CCGT) process of a nearby natural gas fired power plant, resulting in a significantly higher electrical efficiency. This increased efficiency is possible due to the external superheating of the steam to temperatures beyond that possible in a conventional plant. The plant was commissioned in 2004 and is owned and operated by Zabalgarbi S.A. The Spanish engineering company, SENER has the patented rights of the technology (SENER-2 system).

The plant is a conventional single reverse-acting grate incinerator supplied by Martin GmbH, and waste is fed at 30 tonne per hour. The grate is air cooled, with primary combustion air fed under the grate and secondary air introduced higher up the furnace.

The heat from the flue gas is recovered in the boiler and steam is produced at considerably higher pressure, around 106 bar, than in a conventional plant; but the temperature is limited to around 328°C (Unda 2009) in order to prevent excessive corrosion of the superheater tubes that occurs due to the high chlorine content in the MSW feed. This lower steam temperature (most modern plants typically produce steam at 400°C) helps minimise corrosion and hence operational costs and downtime. Instead of sending this superheated steam directly to a steam turbine as in a conventional plant, the steam passes to a second heat recovery boiler.

A 46 MW gas turbine produces electricity from natural gas independently. The hot exhaust gases from this gas turbine contain substantial energy and are passed to the heat recovery boiler raising the temperature of the 100 bar steam from the waste combustion process from 330°C to 540°C (at constant pressure). Corrosion in this boiler is not a problem as the exhaust gases from the gas turbine are free of impurities that can corrode the boiler tubes.

The superheated steam is then passed to a two-stage, 56 MW steam turbine which uses a reheat cycle to further increase efficiency. Steam is expanded through the first high-pressure stage, dropping the pressure to 30 bar. It is then passed back through the heat recovery boiler where it is reheated to 540°C, and passed through the low-pressure stage. Exhaust steam is then condensed in a water cooled condenser which uses water from the nearby Kadagua River. Water cooled condensers are more effective than air cooled condensers at cooling and condensing the steam exiting the turbine, which gives a further slight increase in efficiency over a conventional plant. An on-site water treatment plant cleans the river water to a standard suitable to use at the plant (WSP 2013). The schematic diagram of the Boiler and Flue gas cleaning system is shown in Figure 5.6 (Unda, J 2009).

Flue gas cleaning: The Zabalgarbi plant has a conventional air pollution control system to treat the flue gases: (a) Reduction of NOx utilizing ammonia injection (SNCR) and flue gas recirculation; (b) removal of acid gases (SO_2, HF, HCl) using semi-dry system, spray drier with lime injection; (c) elimination of heavy metals, dioxins/furans, and other pollutants by adsorption via active carbon injection; (d) dust removal using fabric filter. The cleaned flue gases exit the plant via the stack. Flue gases from the gas turbine are rejected via a separate, lower stack. These exhaust gases do not require cleaning as the controlled combustion of natural gas in a turbine releases minimal hazardous or harmful substances given the high purity of the fuel.

Residues: The wastes remaining after combustion of MSW are slag, scrap and ash. Slag (bottom ash) is collected at the bottom of the furnace. It is classified as non-hazardous waste and accounts for 18.31% by weight and 8% by volume of the MSW entering the plant. It is composed of inert materials such as glass, debris and ceramics. It is reused as aggregate material in civil engineering works. Ferrous scrap accounts for 2.28% by weight of the waste introduced to the plant. It is recycled for use in steelworks. Ash is composed of fly ash and residues from gas scrubbing. It represents 3.60% by weight and slightly less than 1% by volume of the MSW delivered. Lime makes up from 50 to 60% of its composition by weight. There are also small

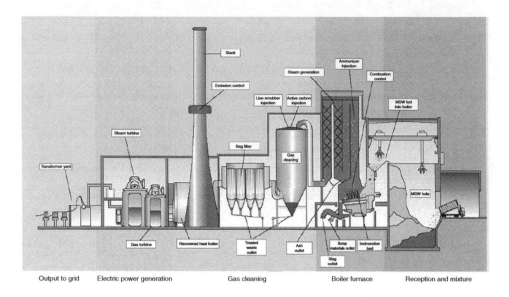

Figure 5.6   SENER-2: Boiler and Flue gas cleaning system schematic (source: Unda, J. 2009; reproduced with the permission of Dr. Juan Unda; www.sener.es).

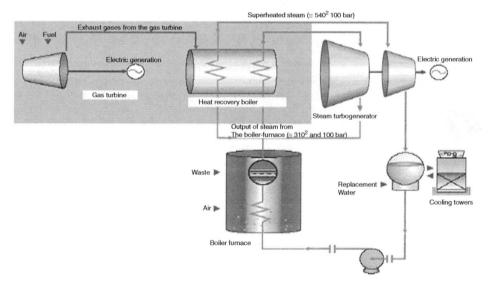

Figure 5.7   Schematic of the steam cycle of the Plant (source: Unda, J. 2009; reproduced with the permission of Dr. Juan Unda; www.sener.es).

metal fractions mixed with the waste: aluminium, iron, magnesium and potassium and trace amounts of Zn (0.75%), Pb (0.17%), Ni (0.0076%), Cd (0.0065%), Cr (0.0139), Hg (<0.001%) and As (0.0013%). This ash has been characterized by an approved technology centre and found non harmful, toxic or mutagenic substances.

Table 5.1   Emissions from the Plant (2008).

Emissions	Zabalgarbi Plant (mg/Nm³)	EU emission limits (mg/Nm³)
NOₓ	147	200
Particulates	3.33	10
HCl	6.85	10
COT	0.86	10
CO	2.34	50
HF	0.097	1
SO₂	12.9	50
Pb + Cr + Cu + Mn + As + Ni + Sb + Co +V	0.0255	0.5
Hg	0.0009	0.05
Cd + Tl	0.011	0.05
Dioxins/Furans	0.0031 ng TEQ/Nm³	0.1 ng TEQ/Nm³

It is stabilised via an inertisation process to prevent leaching of heavy metals, and sent to storage in an authorised facility (Zabalgarbi Plant website).

The plant is capable of processing between 230,000 and 250,000 TPA of MSW from Bilbao. However, the availability in 'normal' operation mode of 7,320 hours per year (83.5%) implies a total of around 220,000 TPA used in the combined cycle process (maximum energy recovery) in practice. The plant is capable of incinerating waste without utilising the combined cycle, but there will be no energy recovery under such circumstances.

The rated gross output of the plant is 99.5MWe. The net output is 95 MWe. The quoted efficiency of the plant varies somewhat according to differing sources (e.g., 47% by Unda 2009), but the net electrical efficiency appears to be around 42% (WSP 2013).

In summary, the Bilbao combustion facility is an example of a modern plant utilising the exhaust heat from an adjacent gas turbine power plant to perform reheating of the steam produced by the heat recovery boiler and operate with a thermal efficiency >40%. (References: Unda, J. 2009; WSP 2013; Zabalgarbi website)

5.4   RIVERSIDE PLANT, BELVEDERE, LONDON, UK

The energy-from-waste plant at Belvedere, London is a river-served waste management facility, owned and operated by Riverside Resource Recovery Ltd. The plant is Hitachi Zosen Inova stoker grate incinerator and the key outputs are about 66 MW electricity generation; and 214,000 tonnes of incinerator bottom ash and metal and 22800 tonnes of APC residue per annum. It has the potential for generating 478,000 MWh/year (Gillett 2011). The energy recovery concept of the facility is also designed for potential off-take of steam or hot water for district heating purposes for future developments.

The schematic of the Plant is shown in Figure 5.8. The waste is delivered to the Plant after several materials are separated for recycling via kerbside recycling schemes

Riverside / UK

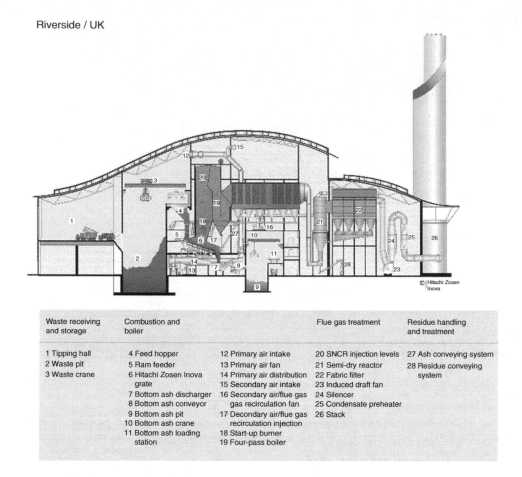

Waste receiving and storage	Combustion and boiler		Flue gas treatment	Residue handling and treatment
1 Tipping hall	4 Feed hopper	12 Primary air intake	20 SNCR injection levels	27 Ash conveying system
2 Waste pit	5 Ram feeder	13 Primary air fan	21 Semi-dry reactor	28 Residue conveying system
3 Waste crane	6 Hitachi Zosen Inova grate	14 Primary air distribution	22 Fabric filter	
	7 Bottom ash discharger	15 Secondary air intake	23 Induced draft fan	
	8 Bottom ash conveyor	16 Secondary air/flue gas recirculation fan	24 Silencer	
	9 Bottom ash pit	17 Decondary air/flue gas recirculation injection	25 Condensate preheater	
	10 Bottom ash crane	18 Start-up burner	26 Stack	
	11 Bottom ash loading station	19 Four-pass boiler		

Figure 5.8 Schematic of the Riverside Plant (reproduced with the permission of Hitachi Zosen Inova AG; photo sent by Dr. Michael Keunecke).

and the Household waste recycling sites. Nearly 85% of the waste (about 536 thousand tonnes) is delivered by barges and only 15% of the waste (about 80 thousand tonnes) is delivered by road. The waste is tipped into a storage bunker through one of 12 tipping bays. Infrared cameras monitor the waste in the bunker for hot spots allowing an automatic fire control system to operate if temperature limits are exceeded. Combustion air for the process is drawn from above the bunker thus minimizing odours and dust in the bunker and tipping area. Two overhead grabbing cranes are used to charge each of the three boilers (four-pass boiler with a thermal capacity of 79.5 MW) with waste, up to a throughput of 30 tonnes per hour each. Waste enters the combustion chamber via a vertical feed chute and is pushed by hydraulic rams from the feeder table onto a four-row Hitachi Zosen Inova stoker grate consisting of alternate rows of fixed and moving cast steel bars, carrying a moving bed of burning waste. The moving grate mixes and agitates the waste to allow an optimal burnout of the diverse waste

Figure 5.9 Photo of the Riverside Resource Recovery EfW facility (reproduced with the permission of Hitachi Zosen Inova AG; photo sent by Dr. Michael Keunecke).

fractions. Complete combustion is achieved by the use of primary air nozzles in the grate bars and secondary air nozzles above the waste bed and varying the speed of the grate and the temperature and flow of air.

The heat value of the waste can vary considerably even after mixing in the bunker. Therefore, controlling the combustion process requires many variable inputs to ensure a stable output together with a combustion process that continuously operates safely within standard environmental limits.

The plant operation data made available are: Maximum through put waste 670,000 TPA; Waste throughput per stream 31.8 t/hour; Steam produced per stream 54 t/hour; Steam pressure 72 bar; Steam temperature 427°C; Flue gas outlet temperature 190°C (Hitachi Zosen Inova brochure, Gillett 2011; Hunt 2011).

The combustion control system has an interface with the Metso DNA system to allow total control of the combustion process. Steam is generated through heat transfer from the combustion process into water filled boiler tubes. Superheated steam from the boilers is used to operate the 72 MW turbo generating unit. Around 6 MW of this is used in the plant and the rest is pumped into the grid. The Metso DNA is a single integrated automation and information platform for process control and it monitors and controls water levels, steam flows, steam temperatures and pressures (Alison Peck). The exhaust steam from the turbine is condensed by a bank of air-cooled condensers. All the operational aspects of the plant are controlled and monitored to the strict standards of the EU Waste Incineration Directive.

Flue gas cleaning: After leaving the horizontal last pass of the boiler and still maintaining sufficient temperature levels for efficient and reliable removal of pollutants, the gas cleaning takes place in the semi-dry system consisting of a reactor in combination with a fabric filter. This well proven technology keeps the plant in safe compliance with the emission limits of the EU by operating below them.

Hydrated lime and activated carbon are used as reagents for the removal of acid gas pollutants, heavy metals and dioxins. Small particles are separated in the fabric filter. The facility is also designed to operate with bicarbonate (as the reagent) in the event that this would prove to be suitable in the future. The NOx-levels are maintained using SNCR by injecting ammonia. Finally, the gas passes through a fabric filter before being released through the stack.

The principal residue from the process is bottom ash. This is collected in the ash bunker and loaded into covered containers and transported on the River Thames to Ballast Phoenix's plant at Tilbury for metal recovery; and for recycling ash for reuse. To start with, the bottom ash is separated into various size fractions with the remaining metals extracted and sent for further processing. The processing stages include washing, screening, crushing and scrubbing. The finished product is now classified as incinerator bottom ash aggregate (IBAA) and is suitable to be used in road building and construction aggregates. The processed IBA is used as fill material, asphalt and foamed asphalt, cement bound materials, lightweight blocks, foamed concrete and pavement concrete. The APC residues are classified as hazardous waste and are taken to a Veolia Environmental treatment facility where they are conditioned and treated, prior to landfill.

The emissions to air are discussed in the Cory's website (http://www.coryenvironme ntal.co.uk/page/rrremissions2012.htm).

In summary (Hitachi Zosen Inova website),
Annual capacity: 585,000 TPY; Number of trains: 3; Throughput per train: 31.8 t/h;
Thermal capacity per train: 79.5 MW;
Waste type: Domestic solid waste; Calorific value of waste: 7.0–13 MJ/kg;
Grate type: Hitachi Zosen Inova grate R-100104; Grate design: 4 rows with 5 zones
    per row;
Grate size Length: 10.25 m, width: 10.40 m; Grate cooling: Air-cooled; Boiler: Type
    Four-pass boiler, horizontal;
Steam quantity per train: 96.5 t/h; Steam parameters: 72 bar/ 427°C;
Flue gas outlet temperature: 190°C;
Flue gas treatment: Concept SNCR, semi-dry system; Flue gas volume per train
    170,000 m³/h; Electric power generation, 65 MW at 100%;
Efficiency: Operates with a relatively high thermal efficiency of 27%;
Residues: bottom ash, 146,250 TPA;
Flue gas treatment: 10,015 TPA;
Waste deliveries and Bottom ash transport: by barges via the river Thames.
(References: Hitachi Zosen Inova website, WSP 2013, Gillett 2011)

5.5 CASE STUDIES FROM JAPAN

Japan can be considered as a leader in developing and implementing traditional and novel thermal conversion technologies. The principal and most popular technology until 1999 was grate firing type incineration. In January 2000, the new strict law, Act on Special Measures against Dioxins, was enacted to reduce drastically the emission of dioxins from incineration plants. As a result, over one hundred thermal treatment plants based on relatively novel processes such as direct smelting (JFE, Nippon Steel), the Ebara fluidisation process, and the Thermoselect gasification and melting process have been developed. These processes have emissions not only as low as the conventional waste combustion process but produce vitrified ash as the law demands because of limited land availability.

Japan generates about 65 million TPA of waste, and treats 40 million tonnes thermally. The rest is being recycled and/or composted and only 2% is landfilled.

As transportation of 'as collected' MSW from one municipality to another is not allowed in Japan, the grate combustion facilities are relatively small. The MSW of several communities is processed to RDF in local RDF facilities and is then transported to a central EfW that serves several communities. Because of the strict rules and very limited land availability, the newer technologies, though seem to be economically unviable in other regions of the world, are established in Japan. However, still 84% of the MSW is processed in grate combustion plants (Themelis and Mussche 2013; Nagayama, S). Some Japanese cities have made their MSW incinerators the center of community complexes with indoor gardens, meeting halls, second-hand shops, and offices of NGOs.

Here, *two case studies* are presented: the first one using conventional grate technology, and the second utilizing the novel gasification and melting process.

(a) Toshima Incineration Plant, Tokyo

The Toshima Incineration Plant is equipped with two 200 TPD atmospheric Bubbling fluidized bed incineration boilers constructed by Ishikawajima-Harima Heavy Industries. FB technology was chosen to reduce the required plant floor area as it was located in a residential zone. Because of the furnace's vertical design, the incineration capacity per unit area is much greater than for a grate fired unit.

The BFB incinerators are designed to operate on waste feed with a heating value, 7.1–13.4 MJ/kg. The energy content of the received waste was 9.4 MJ/kg and contained 5–10% noncombustibles. The waste is combined with combustion air in hot sand under vigorous mixing. There are basically three zones in the vertically oriented incinerator: the fluidized bed, the freeboard and the boiler. At the bottom of the vessel is the dense bed. The fluidizing air enters here through a horizontal tubing grid (distributor) just above the incinerator floor. At a higher elevation in the fluidized bed, primary combustion air (approx. 7,550 Nm³/h) is injected. Temperature in the bed is maintained at about 550–630°C, hot enough to drive off volatiles and fully combust the MSW fed at the top of the bed. If the temperature should rise above 630°C, cooling water sprays are activated automatically. Ash and sand periodically migrate downward and are removed at the incinerator bottom. Sand is separated from the ash, graded, and returned to the top of the dense bed. Each incinerator contains 57 m³ of sand (90 t), some of which is lost as fines with the flue gases, or in the ash stream. It is estimated that periodic make-up results in complete sand change over a period of one to two years.

Above the dense bed is a tall region known as the freeboard where secondary combustion air (approx. 28,800 Nm³/h) is injected at several levels to completely burn off the volatiles. The temperature in this region rises steadily from about 710°C to 1030°C (automatic cooling water sprays are activated should the temperature exceed 1070°C), and gas velocity is such that a residence time (at 850°C) of at least two seconds is achieved for dioxin destruction. In addition to fly ash, some sand fines

Figure 5.10 Photo: Two views of the Plant (source: www.union.tokyo23-seisou.lg.jp.e.de.hp.transer.com/.../toshima/index.ht).

may still be carried by the gases in the freeboard, but these are minimized by prudent velocity control.

The boiler is located above the freeboard. With no combustibles remaining in the gas, and with the aid of cooler air injection, temperatures drop rapidly prior to contact with the boiler tubes (about 480–580°C). This is a natural circulation water-tube boiler, equipped with a superheater. Steam is generated at a maximum rate of 33.3 t/h from each unit, usually at 3.14 MPa (abs) and 300°C. The high-pressure steam is routed to a high pressure steam header, while the flue gases exit the boiler through an economizer to a quick-quench cooling tower (Granatstein and Sano).

Flue gas treatment: Flue gas cleaning begins at the exit of the economizer, where a water spray cooling tower quickly quenches the gases to 150°C, minimizing dioxin formation. At the entrance to the fabric filter, slaked lime and powdered activated carbon are injected into the flue gases to remove heavy metals, dioxins/furans and non-combusted organics, while the baghouse removes particulates. The designed gas treatment rate in the baghouse is about 75000–109000 Nm^3/h (dry).

The flue gases then enter a wet caustic soda scrubbing tower which removes acid gases (HCl, SO$_2$), at a gas treatment rate similar to the baghouse.

Upon exiting the scrubber, the flue gases are dried and heated, by heat exchange with steam generated in the plant, to 210°C before entering the SCR reactor. Here, ammonia is injected into the gas stream as it passes through a honeycomb catalyst to remove NOx.

From the SCR, flue gases enter the 210 m stack (the tallest concrete stack in Japan), containing two flues (one for each incinerator) and an elevator (for maintenance). The inlet temperature to the SCR was chosen to improve the rate of catalytic conversion of NOx (although a temperature of 250–350°C would have been more appropriate), and to ensure an invisible plume emanating from the 210-meter Smokestack.

Energy recovery: Output from the incinerators is 66.6 t/h of steam at 3.14 MPa and 300°C. Enthalpy under these conditions is 2.99 MJ/kg, and total output is 199.1 GJ/h, yielding incinerator combustion efficiency of 89.2% (excellent for fluidized bed furnaces of this size).

A maximum of 58.4 t/h of steam at 2.84 MPa enters the turbine/generator, where 7.8 MWe is generated. Steam at 12 t/h is extracted from the turbine, with 6.5 t/h used in the de-aerator, leaving 5.5 t/h for rating operation. Electrical generation efficiency for the turbo-generator is 16.1%. Waste to electricity efficiency is: $0.161 \times 0.892 = 14.4\%$, a rather low figure by North American standards.

Of the 66.6 t/h of steam generated, 8.2 t/h is directed elsewhere: 5.2 t/h (15.5 GJ/h) is routed to the Health Plaza for heat, air conditioning and pool water heating, and the rest 3.0 t/h (9.0 GJ/h) is used in the plant (e.g., for reheating the flue gases prior to entering the SCR).

The gross efficiency of conversion of energy from waste to end use is therefore: $(28.1 + 15.4 + 9.0 + 15.5)/223 = 30.5\%$. This figure is moderately low, considering part of the heat energy is used directly as steam; however, the allowable steam conditions and energy-intensive environmental treatment are the main reasons. The BFBs perform quite efficiently.

Ash management: Bed ash and fly ash are collected. The bed ash is passed over a permanent magnet drum separator, which removes 0.15% ferrous metals, or 0.6 t/d (design value is 100–150 kg/h). The rest (incombustibles) amounts to 0.35%, or 1.4 t/d (design value is 200–450 kg/h), which is removed to the landfill. The fly ash (containing some unburned carbon), recovered in the fabric filter, is mixed with chelate and water in the ratio of 50:1:15 to prevent heavy metal leaching. The dry fly ash amounts to 1.7%, or 6.7 t/d (design value is 650-980 kg/h), and is also removed (chelate-treated) by truck to the landfill.

In summary, the Toshima Incineration Plant burns 300 TPD of waste and produces electricity, hot water and road materials. The incinerator runs in a dense urban area, reportedly without producing stench.

(b) JFE Gasification & Melting (Recycle) Power Plant, Fukuyama, Hiroshima Prefecture

The Fukuyama Plant is the largest JFE *gasification and melting plant* in the world established in 2004. In Japan, it is almost mandatory to treat MSW in each city or town; in case the volume of MSW is insufficient, a couple of neighboring municipalities jointly build a plant and operate.

The Fukuyama plant is quite unique because the nine cities and villages scattered in different areas produce palletized RDF in their seven units and efficiently transport to this plant for treatment. The features of the Plant are given in the Table 5.2.

The average characteristics of the RDF fuel: moisture content = 3.6–5.5 (wt%); combustible fraction = 80.5–85.1 (wt%); and ash content = 10.6–14.6 (wt%); heat value = 18.2 MJ/kg.

Process description

The JFE Gasifying and Melting Furnace is shown schematically in Figure 5.11. This system is the most compact that processes gasification and melting in a single furnace.

Table 5.2 Features of the Plant.

Capacity	13 t/hr (i.e., 314 TPD); 1 line
Feed stock	Pelletized RDF from MSW
Furnace type	JFE High temperature Gasifying and Direct Melting furnace (vertical shaft furnace)
Energy recovery system	Boiler 90.8 t/h; steam, 450°C/60 bars
Flue gas treatment	Dry type: slaked lime and activated carbon injection, Fabric fliter, de-NOx reactor
Slag treatment	Water-granulation conveyor, Magnetic separator, Slag crusher

Combustibles in wastes will be gasified by the high temperature of the furnace, and non-combustibles will be melted and turned into slag and metals for reuse.

Waste in the RDF form are fed into the furnace with coke (approx. 5% of wastes) and limestone (approx. 3% of wastes) from top of the furnace. The coke plays a role as fire grates that keeps flow path for the gas and slag while functioning as the heat source to prevent cool-down of slag. Limestone helps to form fluid slag that can be easily discharged from the furnace bottom.

Since a reducing atmosphere is maintained in the furnace, hazardous heavy metals are vaporized to the gaseous phase and molten ash is converted to safe slag. The molten ash from furnace bottom is quenched in a water-granulation conveyor to form granulated slag and metals. Besides, the working load for discharge of slag and metals is decreased by the adoption of JFE's original molten ash continuous discharge system.

The schematic of the Plant's process flow is shown in Figure 5.12. The gasifying and melting furnace can be divided into three zones:

Zone 1 (high temperature combustion and melting) is filled with coke which burns with the fixed carbon in the wastes in the presence of oxygen-rich air (approx. 35%) sent through the main tuyere at the lower part of the furnace. Air is introduced into the furnace through the main, secondary and third tuyeres located along the furnace wall. Oxygen-rich air at the specified flow rate is supplied with a blower after the premixing of air collected from the nearby furnace area and the oxygen from the oxygen generator. As the zone temperature exceeds 2,000°C, non-combustibles in the waste are melted and discharged through the slag outlet at the furnace bottom, in a molten state at around 1,600°C. The generated CO_2 is reduced to CO by a Solution Loss Reaction, and CO flows into the upper Zone 2 as gas of about 1,000°C.

In Zone 2 (drying, pyrolysis and gasification), the gas produced at the lower part is partially burned and kept at approx. 700°C with air sent through the secondary tuyere while maintaining a fluidized state of wastes, coke and limestone, charged from the top of the furnace. By this heat, wastes are preheated and thermally decomposed.

In Zone 3, part of produced gas is burned in a high temperature reducing atmosphere above 850°C by air sent through the third tuyere. Maintaining residence time of two seconds or longer enhances pyrolysis of tar and prevents dioxins formation, thus improving the quality of produced gas significantly.

The produced gas is then introduced into the secondary combustion chamber for complete combustion of the gas. The combustion air is supplied from the waste pit that

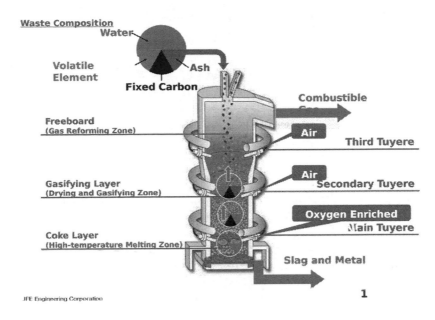

Figure 5.11 Schematic of JFE gasifying and melting furnace (source: JFE; reproduced wih the permission of JFE Engineering Corporation, Daniel Pintado).

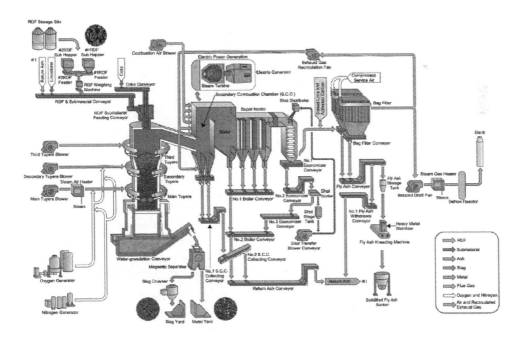

Figure 5.12 Schematic of Fukuyama Plant's Process Flow (source: Nagayama, JFE Technical Report No.16, 2011; reproduced with the permission of JFE Engg. Corporation, Daniel Pintado).

Table 5.3 Emissions analysis (Dry gas at 11% O_2).

Emission	Regulatory value	Analysis value
Dust (mg/Nm^3)	<11	<5.6 (DL)
SOx (mg/Nm^3)	<63	<31.7 (DL)
NOx (mg/Nm^3)	<114	75.2
HCl (mg/Nm^3)	<89	46.7
Dioxins (ng-TEQ/m^{3}N)	<0.06	0.0019

DL: Value is below detection limit (source: JFE Technical report 2011).

also helps to minimize the spread of odour outside the plant. Automatic Combustion Control system regulates the combustion airflow.

Part of flue-gas (low oxygen content) is recirculated to the combustion chamber to keep the temperature stable and ensure high efficiency of combustion.

Steam from the boiler is fed into the steam header via super heaters. All the steam is supplied to steam turbines; and steam extractions from turbines are carried out to improve the power generation efficiency. The JFE's molten slag continuous discharging system helps to decrease the working load.

Plant residues: The data provided for 2009 shows that the RDF throughput was 70753 t; LHV of RDF, 17.8 MJ/kg; coke input, 502 t, and LHV of coke, 29.3 MJ/kg.

The slag produced was 7012 t, and the metals recovered 439 t. All the metals recovered are recycled, and the slag is reused as the backfilling material for road construction.

In Japan, as it is mandatory to melt the ash from the EfW plants, the ash is melt for safety and utilize it as valuable by-product.

Energy recovery: The plant operated at the steam pressure and temperature of 60 bar and 450°C respectively, resulting in an electrical efficiency nearly 27%. The power generated was 10,998.6 MWh.

All the emissions are analyzed to be within the regulated values (Table 5.3). Also, the leaching amounts from the slag measured are found far below the stipulated environmental standards.

Availability: Under Japanese laws, the plants are obliged to treat wastes in their own territories, and the top priority is given to waste treatment and not power generation. In order to keep the plants under continuous operation, any overload to the plants is avoided, and enough maintenance and repair is planned considering 30 years of plant life. As a result, the plants in Japan operate generally from 250 to 280 days per year, and suspend operation for the remaining period for maintenance. In Europe, the plant availability should be more than 8,000 hours (333 days) per year. From a technical point of view, it is possible to increase the availability to 8,000 working hours or more per year even in Japan.

In summary, the JFE High Temperature Gasifying & Melting System is reliable and flexible for treating various kinds of wastes, and highly efficient (power generation efficiency of 27% realized). Since 2003, ten JFE Gasifying & Melting plants (total 20 Lines) are in operation in Japan without any major trouble.

Figure 5.13 Photo of Fukuyama Power Plant (left: RDF production, right: power plant) (produced with the permission of JFE Engg. Corporation, Daniel Pintado).

5.6 ALLINGTON ENERGY FROM WASTE FACILITY, KENT, UK

The EfW facility at Allington, Kent is the largest Fluidized Bed Combustion Plant outside of Japan. The Plant is designed to handle up to 500,000 TPA of non-hazardous waste from households and businesses in and around Kent for energy recovery, and upto 65,000 TPA of source separated recyclables such as paper, card, metals and plastics.

The combustion process is based on rotating fluidised bed technology (ROWITEC) licensed from Ebara by Lurgi Lentjes with the possible injection of urea into the freeboard for NOx reduction. A flue gas cleaning system employing the CIRCOCLEAN technology (explained in Chapter 3) is utilized. The facility consists of three combustion lines.

The internal geometry at the air distribution part of the fluidised bed produces a highly turbulent toroidal mixing regime because Ebara has utiized the advantages of the BFB and CFB designs to produce an optimised design of Twin interchanging fluidised bed (TIF bed) (Figs. 5.14 & 5.15).

The MSW treated at the facility is collected from households and premises where recyclable materials is already segregated. Additional segregation of ferrous metals is done at the site following shredding before the waste is introduced into the combustion chamber. The calorific value of the waste ranges between 6.5 and 12.5 MJ/kg.

1. Waste Feeder
2. Revolving Fluidised Bed
3. Fluidising Air
4. Flue Gas
5. Deflector Plate
6. Non-combustibles
   Discharge Chute
7. Inclined Nozzle Plate

Figure 5.14 ROWITEC FB Firing process Principle (source: Lischke, Gunnar: Lentjes GmbH).

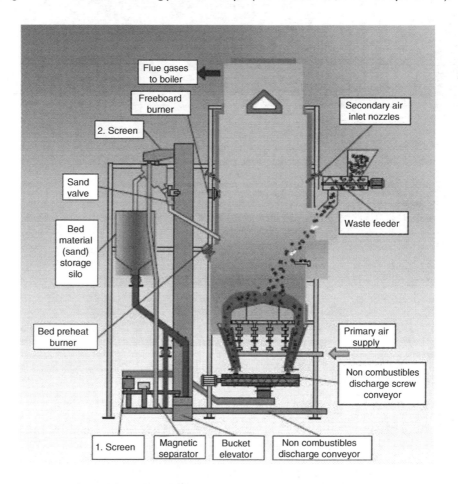

Figure 5.15 ROWITEC Fluidized Bed (source: Gunnar Lischke).

Power and Heat Generation: The plant is designed to generate approximately 43 MW of electricity of which 34 MW is exported to the local supply network. The plant also has the capacity to generate a thermal capacity per line of 53.8 MW.

Air pollution control

The three incineration lines have the facility to inject urea into the afterburner chamber to reduce NOx emissions. ESPs are used for pre-separation of fly ash downstream of the horizontal three-pass boiler. The flue gas cleaning plant with CircoClean technology (discussed in Ch. 3) ensures the compliance of the stipulated emissions limits. The plant is fully compliant with WID of European Union. The monthly Continuous Emissions Monitors reports are available at http://www.kentenviropower.co.uk/enviropower.asp?ID=59.

Ash Handling and Processing: Input wastes up to 25% by weight will reduce to ash during incineration which requires further treatment. Approximately 60,000 TPA of bottom ash is generated which is recycled by a specialist contractor. Air pollution control residues and filters are treated and disposed of in licensed landfill sites.

Availability: This plant experienced technical issues, particularly relating to the refractory lining of the FB combustor and the flue gas cleaning system. The initial operation in 2009 achieved availability of only 65–70% (Tolvik 2011; WSP 2013).

The plant has a very low building profile because most of the fluidised bed combustors and boilers have been sunk 30 metres into the ground.

In summary, the Allington plant is the largest Fluidized Bed Incineration Plant outside Japan utilizting the twin interchanging fluidized bed design. It has a low building profile as the plant is sunk nearly 30 meters into the ground. (Sources: Gunnar Lischke of Lentjes, Tolvik 2011, WSP 2013).

5.7 CHP WASTE-TO-ENERGY PLANT, RENO-NORD, AALBORG, DENMARK

The Plant, built in 1981 to generate and supply thermal energy to the district heating system of Aalborg Fjernvarmeforsyning from waste, is equipped with two Vølund rotary kiln furnace systems. Each furnace line had a combustion capacity of 8 metric tons of waste/hr. In 1991, a combined heat and power line which has a throughput of 12.5 metric tons/hr was added to provide electricity and heat. This line 3 is fitted with a BS-W combustion grate. In 2006, another incineration line (line 4) for combined heat and power generation with a rated throughput of 20 metric tons/hr with a lower calorific value of 12 MJ/kg and an annual volume of 160,000 metric tons of waste was installed. This line 4 replaced the original lines 1 and 2. The line 3 was kept as a replacement.

The line 4 was planned to be a conventional, grate-fired unit with a steam boiler having three vertical radiant flue gas passes and one horizontal convection pass, consisting of an evaporation tube bank, superheater and economiser. The steam data specified was 50 bar/425°C, and feedwater temperature of 130°C.

Process: The waste is fed to the grate by a hydraulic waste feeder at a variable speed corresponding to the energy generation. The continuous, slow forward motion of the waste feeder results in a consistent, continuous flow of waste supplied to the grate. The waste feeder is located at the bottom end of the feed chute, where it feeds the waste slowly to the first section of the grate. The waste feeder consists of several feeder rams; and each ram is driven by a hydraulic cylinder. Each feed ram is air-cooled and has two cooling air inlets to ensure adequate cooling of the ram. The plant is equipped with an air-cooled B&W Volund's DynaGrate®, suitable for conversion to water cooling, if required at a later stage; and with a pusher for feeding of waste that ensures homogenous feeding without the risk of backfire. The grate, which consists of two parallel grate lanes each with a width of 4.4 metres, is divided lengthwise into four sections. There is no limit to the number of grate sections that can be combined, and thus no limit to the length of the grate. Due to good mechanical rotation of the waste on the grate, all of the waste material is exposed to radiant heat from the combustion chamber and combustion air. The efficient distribution of the combustion air ensures controlled and effective combustion that yields very low CO and TOC quantities at the steam generator outlet and stable energy generation. As the primary air supply for this grate is close to the stoichiometric ratio, this result in low temperatures in the waste layer on the grate, which in turn reduces slag agglomeration and minimises the accumulation of encapsulated residues of combustible components. This yields good combustion characteristics and finely granulated slag.

The four-pass boiler has three radiant passes and a horizontal pass with an evaporation heating surface in front of the superheater and an economizer, and is designed to operate with natural circulation. The radiant passes are implemented as a top supported assembly, while the convection pass with the superheater and economiser is implemented as a bottom supported assembly. The boiler is conservatively designed with a large heating surface in order to achieve long-term continuous operating periods without undesirable deposition of ashes during the operating period. The design of Reno Nord boiler was such that the system can be operated with an excess air factor of 1.8, which corresponds to an oxygen concentration in the flue gas of 9.3% by volume (dry).

The turbine system supplied by B+V Industrietechnik GmbH is designed for a swallowing capacity of 120% of the nominal steam generation. The rated power of the generator is 17.918 MW. The turbine system is fitted with a bypass system for receiving the full volume of steam. The schematic of the system is shown in Figure 5.16 with details of temperatures at different stages of the process.

The plant is equipped with a water-cooled wear zone that reduces refractory lining and thereby the build-up of slag. This in turn reduces maintenance costs. The water-cooled wear zone has many advantages: It can absorb approx. 80–120 kW per m² and can receive 1.5–2.8 MW's of extra energy, which is absorbed as radiant heat.

(a)   There are no slag deposits as the wear zone is a relatively cold at 200–300°C. The system therefore retains its active grate area throughout the operational period;

(b)   Operational stoppages for slag removal are not required. Heat absorption in the wear zone is five to ten times higher than in the boiler's two radiation passes;

(c)   Heat absorption in the wear zone reduces the furnace temperature and therefore allows increased heating value and maintains waste capacity;

(d)   The water-cooled wear zone replaces the refractory lining in the system's most heavily used area because a water-cooled wear zone has been proved to have a longer life than refractory lining;

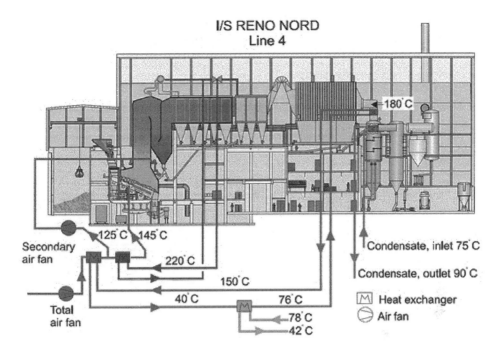

Figure 5.16 The schematic of the Plant design (source: B&W Volund).

(e)   The water-cooled wear zone works actively as an expansion segment between boiler and grate.

The water-cooled wear zone is directly connected to the boiler circuit enabling increase of the overall efficiency of the plant. Due to the high operational temperature, the wear zone is covered with Inconel®.

Flue gas cleaning: The plant is equipped with a modern flue gas cleaning system and hence comply with the EU directive on waste incineration. The cleaning system has a three-section electrostatic filter supplied by Alstom followed by a three-stage wet flue gas cleaning system supplied by LAB S.A including a wastewater treatment plant for the effluent. The dust concentration after the filter is $10 \, mg/Nm^3$ with a flue gas oxygen content of 11% by volume (dry), adequately low at the inlet to the scrubber.

The LAB flue gas cleaning system employs a process that remove ammonia, HCl, SO_2, HF, Hg, heavy metals, dioxins and solid particles from the flue gases of waste incinerators. The flue gas cleaning system also includes a steam-driven ammonia stripper to remove ammonia from the wastewater before it is fed to the wastewater treatment plant. The stripper was supplied by Rauschert Verfahrenstechnik GmbH. SNCR is added for NO_x removal.

In order to achieve high electrical and thermal efficiency, two enamelled and PFA-coated coolers are fitted after the electrostatic filter. They reduce the temperature of the flue gas from 180°C to 90°C before the flue gas enters the wet flue gas treatment plant. The energy extracted is used to preheat the condensate.

Table 5.4 Process performance data.

Process parameter	Guarantee values	300-hour test	Unit
Waste capacity	20	21.72	t/h
Calorific value (LCV)	12	11.28	MJ/kg
Steam production	22.42	22.55	Kg/s
Steam temperature	425	423	°C
Steam pressure	50	48.6	Bar
Input efficiency	66.6	67.9	MW
Electricity produced	17.918	18.232	MW
Thermal efficiency	85.56	87.1	%
Electrical efficiency	26.88	26.93	%
TOC, bottom ash	<20	<0.23	%
Flue gas temp. before superheater	620	530	°C
Outlet temp. boiler	180	181	°C

(Source: B&W Volund, Plant Fact sheet).

Table 5.5 Flue gas emissions (unit: mg/Nm^3; *ng I-TEQ/Nm^3).

Component	Measured value	Guarantee value
Particulate matter:		
(a) After ES filter:	1.6	10
(b) In chimney:	0.21	10
HCl	0	5
HF	<0.1	2
SO_2	7.4	20
Hg	0.0028	0.05
Cd and Tl	<0.0001	0.05
Total Sb, As, Pb, Cr, Co, Cu, Mn, Ni and V	<0.009	0.5
CO	Below detection	50
NO_x	150	200
TOC	Below detection	10
Dioxin and furan*	0.013	0.1

Since the district heating water is returned at a temperature lower than the flue gas temperature after the scrubbers, this difference in temperature can be exploited for further production of district heating. In this process such a large part of the water vapour content of the flue gas is condensed that the unit becomes self sufficient in water for both the wet flue gas treatment and the cooling of the bottom ash. It means, the part of the waste that consists of water is *recycled*. This also entails an exploitation of nearly 100% of the lower calorific value of the waste.

Performance: The nominal capacity of the plant is 20 t/h waste at a calorific value of 12 MJ/kg, corresponding to approx. 160,000 TPA. The boiler generates 80 t/h steam at 50 bar, 425°C. The turbine generates approx. 18 MW of electricity, pumped into the main grid. Besides, the plant supplies approx. 43 MW of heat to the district heating network in Aalborg. The efficiency is approx. 100%, and the energy produced supports about 16,000 houses with electricity, and about 30,000 houses with district heating. Table 5.4 provides test data which refers to 'before' flue gas cleaning.

During the 300-hour test, bottom ash and ash samples were also taken within 12 hours for TOC analysis. The flue gas emissions were also measured; the measured values and the guarantee values are compared in Table 5.5. The values are well below the guarantee values. The wastewater is treated to the point that they fulfill the statutory requirements for direct discharge into the nearby water body.

Bottom ash treatment: Around 25% of the waste incinerated at the Plant becomes bottom ash and is composed of 98% inorganic material and about 2% is non-combusted organic material.

The bottom ash is then further processed to separate the ferrous and non-ferrous metal from the mineral fractions. The ferrous and non-ferrous metals are recycled and the mineral fractions can be used for sub-base layers in road buildings, paths and foundations.

In summary, the performance of the Plant (Line 4) after approximately one year of operation confirms that the main focus in the construction of the Plant, namely, the use of safe, reliable and proven technology in combination with high energy efficiency and long continuous operating intervals, while complying with the stringent environmental requirements specified for Line 4 has been fulfilled. This Plant is considered to represent BAT (Best Available Technique) in Denmark.

5.8   WASTE INCINERATION PLANT, ARNOLDSTEIN, AUSTRIA

The fluidised bed reactor of the waste incineration plant at Arnoldstein is operated by Asamer Becker Recycling Gesellschaft mbH continuously since January 2001. In 2001, 26,000 tons of hazardous and non hazardous wastes (oily waste, solvent-water mixtures, treated and untreated wood waste, wood packaging, plastic waste, sludge and waste water) were combusted. The average calorific value of the waste was 5–30 MJ/kg; the thermal output, 8 MW, and operating hours (test operation) were 7.300 hrs.

Process: The process flow diagram is shown in Figure 5.17. The plant basically consists of the following units: Treatment hall for crushing and grinding and mixing of wastes and space for intermediate storage of wastes; Firing system consisting of stationary fluidised bed reactor with waste heat boiler; Flue gas cleaning system comprising electrostatic precipitator, two-stage wet scrubbing with NaOH scrubber, flow injection process and catalytic flue gas cleaning system (clean gas application), and central waste water treatment plant. Oil is used as additional fuel for start up and shut down when necessary.

Mixed, crushed and grinded solid wastes are fed into the bunker by means of a crane. The bottom of this bunker is constructed as slowly moving conveyer belt. Waste discharged from the bunker falls onto another conveyor belt and is conveyed into a charging hopper for a dosing screw. The dosing screw charges the solid wastes regularly onto a so-called throw feeder, which distributes the waste uniformly across the fluidised bed. Liquid wastes are injected by means of a lance. For start up of the plant two oil burners are installed. The combust air which is the exhaust air from the waste storage facility and the tanks is introduced into the combustion chamber as secondary air through nozzles as conveying air for recirculated bed ashs. A control system for the regulation of the above is installed.

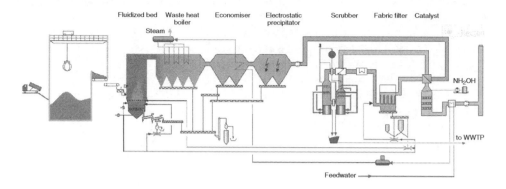

Figure 5.17 Process flow diagram of the Plant (source: State of The Art for Waste Incineration Plants: Federal Ministry of Agriculture and Forestry, Environment and Water Management, November 2002).

The combustion chamber is a brick-lined stationary fluidised bed system. Immediately above the reactor the afterburning zone with secondary air injection is arranged. Hydrated lime and limestone from the flow injection process are pneumatically conveyed into the combustion chamber for pre-separation of SO_2.

The waste heat boiler is a horizontal type with radiation heating surfaces in the first and convection surfaces in the second part. The boiler is followed by a feedwater preheater. The saturated steam produced in the system is fed into the local steam network using a pressure reducing valve.

Flue gas cleaning system

Dedusting: Dedusting of flue gases leaving the boiler is done by an electrostatic precipitator. The temperature of flue gases entering the ESP depends on the boiler load and the travel time.

Wet flue gas cleaning: The wet flue gas cleaning system consists of a co-current scrubber with acid circulation water and a counter current scrubber with NaOH as neutralizing agent. Each scrubber is followed by a droplet separator. Heat from the flue gases entering the scrubber is transferred to the flue gases leaving the scrubber by means of a gas/gas heat exchanger. The outlet temperature can be regulated by a downstream steam heated gas preheater.

Flow injection process: The flow injection unit consists of a flue gas channel with injection of furnace coke, limestone and hydrated lime and a fabric filter. The operating temperature is about 120°C. The added chemicals are recirculated several times and then injected into the combustion chamber.

Catalytic flue gas cleaning: It exclusively serves as NOx reducer. An aqueous solution of ammonia (25%) is used as reducing agent. The exiting flue gases are cooled in a heat exchanger, and the heat is used for preheating water for the feedwater tank.

The emissions from the Plant with reference to 2001 is given in the Table.

Parameter	Emission [mg nm^{-3}]^a	Total mass [kg yr^{-1}]^{b,d}	Specific emissions [g t^{-1}]^{c,d}
Dust*	1,5	169.73	6.53
HCl	0.14	15.84	0.61
HF	0.038	4.3	0.16
SO$_2^*$	<5	565.75	21.76
C$_{org}^*$	<1	113.15	4.35
CO*	<5	565.75	21.76
NO$_X$ as NO$_2^*$	<150	16,972.5	652.79
Cd	0.003	0.34	0.013
Hg	0.003	0.34	0.013
PCDD+PCDF	0.022 ng/Nm^{-3}	2.489 mg yr^{-1}	0.096 µg t^{-1}

Continuous measurement

a. *Half hourly average values in mg N/m^3; dioxin emissions are given in ng N/m^3 (11% O$_2$; dry flue gas; standard conditions)*
b. *In kg/yr, dioxin loads in mg/yr*
c. *Emissions related to one ton used waste in µg t^{-1}; dioxin emissions in µg t^{-1}*
d. *Total mass and specific emissions are calculated based on average half hourly mean values, using the quantity of dry flue gas (5,388 Nm^{3}t^{-1} waste; calculated from the hourly flue gas volume of 15,500 Nm^3, the operating hours and the waste input) and the waste quantity (26,000 t/yr).*

Output flows (ref. Year, 2001) (WERNER, 2002):

Steam (25 bar; 180°C): 4.5 t/h; Ash: 9,000 TPA; Ferrous scrap: 170 TPA; Filter cake: 200 TPA; Waste water: 13,000 m^3/yr; Flue gas: 15,500 Nm^3/hr

The waste water is treated in a plant using heavy metal precipitation, neutralisation and gypsum precipitation; and then released into the receiving water.

The residual wastes (bottom ash, fly ash etc.) were tested and found the concentration of pollutants was far below the regulated limits.

(Source: State of the Art for Waste Incineration plants, Published by Federal Ministry of Agriculture and Forestry, Environment and Water Management, Devision VI/3, Stubenbastei 5, 1010 Wien, Austria, November 2002; Authors: Josef Stubenvoll (TBU), Siegmund Böhmer (UBA), Ilona Szednyj (UBA); Coordination of the joint study: Gabriele Zehetner)

5.9 THE MAINZ WASTE-TO-ENERGY PLANT, GERMANY

Considered as innovative in design in many respects, this waste-to-energy plant has been operating in Mainz since 2003. The steam produced in the plant is released into a highly efficient combined-cycle power plant (CCPP) for generating power, thus replacing the primary energy source, natural gas, and reducing CO$_2$ emissions. The process technology used is designed to minimize the environmental impact such that the emissions are well below the stipulated limits.

The MARTIN reverse-acting grates installed in the Mainz WTE plant facilitate a flexible response to market- induced changes in waste composition.

Waste reception: The vehicles which bring the waste unload the waste in a sealed tipping hall via one of 7 tipping points into the refuse pit. Bulky waste is shredded by rotary shears separately and transported to the refuse pit on a conveyor belt. Two

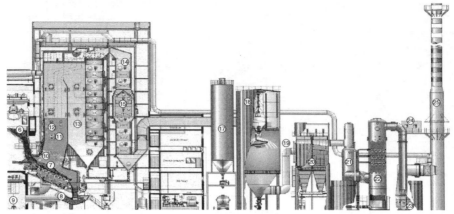

1 Tipping hall, 2 Crushing bulky waste, 3 Refuse pit, 4 Waste crane, 5 Suction facility (not shown in the figure); 6 Feed hopper, 7 Martin Reverse-acting grate and furnace, 8 Martin Discharger, 9 Bottom ash bunker with ash-handling crane, 10 Overfire air & injection system for thin sludge, 11 Auxiliary and start up burners, 12 SNCR NOx reduction, 13 Steam generator, 14 soot blowers, 15 slip catalytic converster, 16 Turbine/generator for in-plant requirements, 17 Dust sili, 18 Spray absorber, 19 Injection of adsorbent agent, 20 fabric filter, 21 Pre-scrubber, 22 Scrubber, 23 Induced draught fan, 24 emission monitoring area, 25 Stack

Figure 5.18 Schematic of the Plant lay out (source: www.mhkw-mainz.de/..).

crane units installed in the refuse pit and intelligent waste management guarantee homogenous and constant combustion of residual waste.

The required combustion air is extracted from the refuse pit by suction, thereby generating negative pressure and preventing odour emissions from escaping into the tipping hall and into the ambient air. The air is preheated by heat exchangers and injected into the furnace. The combustion temperature exceeds 1000°C.

Thermal treatment: The residual waste which passes though the feed chutes of the three combustion lines is conveyed to the combustion grates by means of feed rams. Gas burners are used to start the respective combustion lines. These preheat the combustion chamber to the minimum required combustion temperature of 850°C. Since waste has a greater heating value, combustion takes place independently after the gas burners have been switched off and without any additional supply of primary energy. The gas burners are very rarely switched on in order maintain the minimum combustion temperature. The plant lay-out is shown in Figure 5.18.

The combustion residues remaining on the grate after a residence time of approx. 1 hour are transported to the bottom ash bunker via a wet-type discharger. The remaining bottom ash is treated externally in a bottom ash treatment unit in a series of process steps which separate the metal from the mineral fractions. Iron scrap and non-ferrous metals are recycled for use in the ironworks industry and the mineral fraction is used in landfills and road construction as a substitute material instead of using new products. Thus even the bottom ash is recycled.

The heat released during waste combustion is converted to steam in 4-pass vertical steam generators, and a small proportion is converted to electricity in a turbine to serve in-plant requirements.

This facility is designed to generate steam only. The conversion of the steam to electrical energy is carried out in the neighboring 400 MW combined cycle power plant (CCPP) owned by Mainz-Wiesbaden AG.

System details: The Entsorgungsgesellschaft Mainz mbH facility utilises three MARTIN Reverse-acting grates of which lines 1 and 2 were commissioned in 2003 and line 3 was commissioned in 2008.

The initial capacity of the facility with lines 1 and 2 was 230,000 TPA, which was increased by approximately 50% to around 340,000 tonnes with the introduction of the third line.

(a)   3 MARTIN Reverse acting grates; Runs = 3; Number of steps = 13; Run width = 1875 mm; Grate width: Lines 1 & 2 – 5945 mm; Line 3 – 6320 mm

(b)   Technical data:
Waste throughput: 3×16.2 t/h at heating value = 9.8 MJ/kg
Gross Heat Release: 1 & 2 lines: 2×44 MW; Line 3: 48 MW
Steam Pressure – 42.3 bar
Steam temperature: 1 & 2 lines – 400°C, Line 3 – 420°C
Flue gas temperature: 190–230°C
The process exhaust gases are emitted to atmosphere via a 95 m stack.

Flue gas cleaning: The flue gas cleaning system consists of the following stages:

(a)   SNCR unit with aqueous ammonia injection into the first boiler pass above the furnace in order to reduce the NO_x emissions while forming nitrogen gas and water vapour.

(b)   High-dust catalytic converter to reduce the surplus ammonia that has not reacted.

(c)   Spray absorber with lime milk added via an atomizer wheel to reduce the temperature and pre-separate acidic gases such as SO_2, HCl and HF.

(d)   Facility for adding activated coke upstream of the fabric filter to bind dioxins/furans, heavy metals and other pollutants.

(e)   Fabric filter for separating dust.

(f)   Pre-scrubber with water injection to reduce acidic pollutants.

(g)   Main scrubber with lime additive to remove residual flue gas components as well as mercury.

The flue gas is saturated with water vapour by the two scrubbing stages. This is why a plume of water vapour can always be seen at the opening of the stack, typically indicating that "flue gas cleaning" is in progress.

Exhaust gases exited to atmosphere from the Plant are subject to continuous and discontinuous measurement to ensure compliance with §18 (public information) of the 17th Publish BimSchV and are published on the facilities website and given in Table 5.6 (http://www.mhkwmainz.de/anlagentechnik/messwerte.php).

The half hourly and daily mean values are also transmitted to the competent licensing authority. The summary of the emission limited in 17th BimSchV and the permitted limits for the facility are given.

Table 5.6 Actual permitted and actual (2011) emission levels for the plant (continuous measurement) (unit: mg/m³).

Pollutant	17th BImSchV	Permitted limits	Yearly Ave. Line 1	Yearly Ave. Line 2	Yearly Ave. Line 3
CO	50	50	4.31	5.47	5.39
Total dust	10	8	0.17	0.30	<0.01
Total carbon	10	10	0.34	0.24	0.01
HCl	10	8	<0.01	0.54	0.44
SO$_2$	50	50	5.89	3.90	5.00
NO$_x$	200	150	114.22	117.41	117.42
Mercury	0.03	0.03	0.001	0.001	0.001
Ammonia	–	10	0.44	0.36	0.64
Cd and Th	0.05	0.05	0.0003	0.0002	0.001
Total Sb + As + Pb + Cr + Co + Cu + Mn + Ni + V + Sn	0.5	0.5	0.028	0.03	0.04
Total As, Benzo-Pyrene, Cr (VI)	0.05	0.05	0.002	0.006	0.001
Dioxins/Furans (ng/Nm³)	0.1	0.08	0.002	0.002	0.005

The Plant achieved values that fall significantly below these limits. Continuous and intermittent measurements ensure that emission values are monitored and controlled at all times.

Ash handling and processing

Approximately 26% of the input materials become slag which requires further treatment. The bottom ash is treated externally in a bottom ash treatment unit in a series of steps which separate the metal from the mineral factions. Iron scrap and non-ferrous metals are recycled for use in the ironworks industry and the mineral fraction is used in landfill and road construction as substitute materials for virgin aggregates. Air Pollution Control residues are used for infilling old salt mines. The material flows through the process are shown in Figure 5.19.

Energy Production: The Gross Heat Release is 2 × 44 MW from Lines 1 and 2; and 48 MW from Line 3. Most of the steam produced in the facility is exported at 40 bar/ 400°C to the neighboring 400 MW CCP Plant where the steam is superheated to 550°C without using additional primary energy sources and is converted to electricity in a highly efficient manner using a gas turbine. Based on data provided by MARTIN, the net efficiency of this facility is >40% (WSP 2013).

> *In summary, the Mainz WtE plant is an example of a modern German plant producing high efficiency power in combination with the neighbouring CCPP and meeting stringent emission limits. The plant also achieves high availability.* (Source: The Mainz Waste-to-Energy Plant, published by Entsorgungsgesellschaft Mainz mbH, Gaßnerallee 33, 55120 Mainz, Germany, December 2007; available at http://www.mhkw-mainz.de/fileadmin/downloads/EGM_infobro3_eng_20.pdf; & WSP 2013.

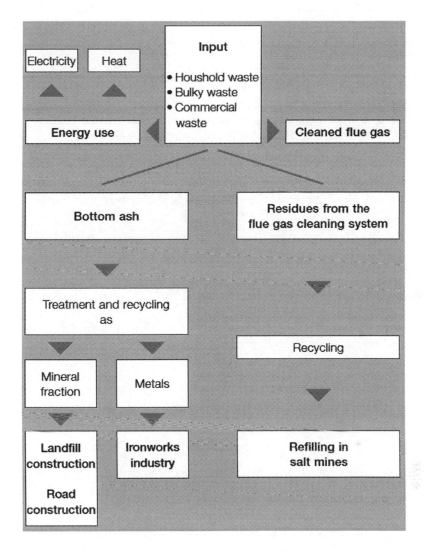

Figure 5.19 Representation of materials flows in the Plant (source: www.mhkw-mainz.de/..).

5.10 INDAVER INTEGRATED MSW TREATMENT PLANT, DOEL (NEAR ANTWERP), BELGIUM

The objective is an *integrated concept* with maximum energy and materials recovery, and an all-out attention to environment and safety.

The state-of-the-art grate furnaces with energy recovery and flue gas treatment followed by wet bottom ash treatment are used by Indaver, the operator of the Plant. Due to the stringent environmental regulations, several modifications have been applied to optimize and intensify the incineration process such as addition of a wet flue gas cleaning to the half-wet flue gas cleaning, zero water discharge, and high energy recovery.

In order to recycle the large quantities of bottom ashes, a 'wet bottom ash treatment' plant was developed and included.

Plant operation: The incineration plant is comprised of three grate furnace lines, i.e., a furnace with a boiler, gas washing and a chimney. Two lines each have a nominal treatment capacity of 13.3 ton/h, the third one of 21.5 ton/h, in total 48.1 ton/h. Figure 5.20 schematically represents the components of a grate furnace line. The waste is delivered via a bunker and brought from the bunker onto the moving grate using a crane. Here it is incinerated at a minimum temperature of 850°C (up to 1000°C) generating bottom ash (incinerator ash) and flue gas.

Each line is equipped with a boiler with a thermal power of 40.0 (two lines) or 67.5 MW (one line) depending on the capacity (total thermal power 147.5 MW) and the flue gas is passed through the boilers. These boilers are equipped with an economiser in order to heat the feed-water and with a superheater to heat the saturated steam (45 bar, 257°C) further to 40 bar and 400°C. During cooling of the flue gas in the boiler, boiler ash is deposited and is collected.

Flue gas treatment: In the spray-dryer, lime milk (slurry of $Ca(OH)_2$) is injected. The flue gases are cooled from 230 to 160°C by the vaporisation of water from the injected lime milk. Activated carbon is injected immediately behind the scrubber to adsorb dioxins, PAHs and heavy metals such as Hg. The activated carbon, along with salts and particles, is retained on the baghouse filter as flue gas cleaning residue. In the wet scrubber placed after the baghouse filter, the flue gas is further cooled to 60°C and washed intensely by sprinkling slurries containing limestone ($CaCO_3$) and lime, $Ca(OH)_2$ at different levels.

Acidic compounds, HCl and SO_2, are converted by reaction with the lime and limestone to non-hazardous compounds ($CaCl_2$ and CaSO). The effluent from the wet scrubbers is treated by filtration, using a belt-filter, in order to separate, e.g., gypsum. The water is then recycled into the spray dryer, so that all water leaves through the chimney.

Bottom ash treatment: Figure 5.21 schematically shows the wet treatment of bottom ash consisting of washing, sieving and separating. After removing large parts of metal and stone, material >50 mm is then removed by a screening and washing installation using water. Ferrous metals are removed magnetically and the rest of the material >50 mm is partly sent back to the grate incinerator. The finer material (<50 mm) passes through a washer barrel to separate from granulates, the light organic material, which is also sent back to the grate incinerator. Another screening and washing section separates the particles in three different fractions: 6–50 mm, 2–6 mm and <2 mm. The ferrous separators retrieve the iron from the two larger fractions and non-ferrous separators, based on Eddy current, separate mainly aluminium and copper. In practice they are always in use for the 6–50 mm fraction, and are optional (only used to improve the quality of the end-product, when the end-product is applied in building materials) for the 2–6 mm fraction, where the yield of non-ferrous metal is low. Finally the fraction <2 mm is separated in a sludge fraction (<0.1 mm) and a sand fraction (0.1–2 mm). The sludge is landfilled without stabilisation. This ash treatment facility has a capacity of 165,000 TPA (OVAM, 2001). Although the installation uses large quantities of water (L/S = 0.5–1), not much water is actually 'consumed' and no wastewater is discharged.

In order to decrease metal leaching, the obtained granulates (6–50 and 2–6 mm) are aged for 3 months in 5–10 m high heaps, located at the site of the incineration plant

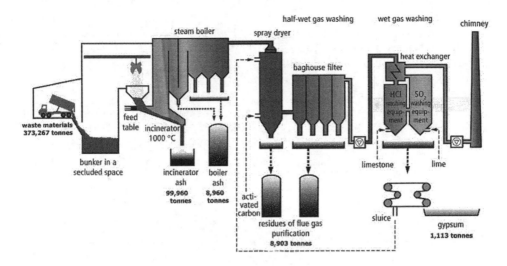

Figure 5.20 Schematic representation of the Grate incineration plant and flue gas treatment (source: Vandecasteele et al. 2007; reproduced with the permission of Prof. Carlo Vandecasteele).

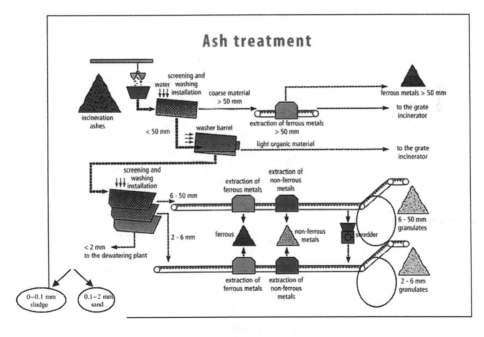

Figure 5.21 Schematic of wet treatment of bottom ash (source: Vandecasteele 2007; reproduced with the permission of Prof. Carlo Vandecasteele).

in open air on a paved floor with collection of the percolating rain water, exposing them to wind and rain. The reactions taking place during this time tend to decrease heavy metal leaching (Meima and Comans 1999; Chimenos et al. 2000; Freyssinet et al. 2002; Jaspers 2002; Van Gerven et al. 2005 a & b).

Treatment of boiler ash and flue gas cleaning residue: Boiler ash and flue gas cleaning residue (scrubber residue, gypsum) are collected in silos, are kept separated from the other ashes and are treated by stabilisation/solidification.

HCl is added to neutral pH and subsequently cement (0.15 ton/ton) is added in order to obtain a monolithic end-product, in which the heavy metals are immobilised (Conner 1990; Gougar et al. 1996; Geysen et al. 2003). The end-product is landfilled in the class 1 landfill for hazardous inorganic waste, operated by Indaver N.V., at Stabroek, close to Antwerp.

Energy recovery

150 ton/h of steam (400°C, 40 bar) are produced corresponding to a net thermal boiler yield of 80%. The steam produced is partly used in the installation itself (ca. 79.10^3 ton of steam per year, 8.9%) or is supplied to an energy supplier (ca. $464 \cdot 10^3$ ton of steam per year, 38.6%) for distribution to industrial companies. The remaining steam is fed to a turbine-driven electrical generator (installed max. power, 21 MW$_e$) that supplies electricity to a utility.

The gross electrical yield of this part of the installation is ca. 24% (net yield, i.e., after deduction of the electricity used in the installation, of 21%). The heat generated in the combustion process is thus made available through two carriers: steam and electricity.

The gross energy conversion yield (eficiency with which the energy content of the waste is converted into useful energy, irrespective of the carrier) is 50.2%, corresponding to a net value of 40.8%. This energy conversion yield may be considered outstanding, as the energy conversion yield of a grate furnace installation is influenced negatively by certain factors (Vandecasteele et al. 2007).

Emissions to air

The installation was shown to present low occupational health risks. The three major pollutants, dioxins/furans, Hg and iodine-131, are all removed to a very large extent and are concentrated in a residual waste stream (boiler ash and flue gas cleaning residue). These streams only amount to ca. 6% (wt) of the incoming waste. After immobilisation by solidification/stabilisation, the end-product, a stabilised monolith, is safely disposed of in a class 1 landfill for inorganic hazardous waste.

> *In summary, it was demonstrated that the integrated installation at Doel, Antwerp, using a grate furnace and wet bottom ash treatment, has a good environmental performance and may play an important role in sustainable waste management.* (Source of the material: Vandecasteele et al. 2007).

What is the best available technique for incineration of waste?

According to technology developers, Heron Kleis of B&W Volund and Soren Dalager of Ramboll (2004), mass burn incineration on a grate or a combination of a grate and a rotary kiln can currently be considered BAT, provided that computerized fluid dynamics (CFD) calculations are used in the design process.

REFERENCES

Andersen, H.B. *BWV Combustion Technology for Generating Energy from Waste—Case Study Reno Nord, Line 4, A High-Efficiency Waste Incineration Plant.*

Chang, M. & Huang, C. (2002) Characteristics of chlorine and carbon flow in two municipal waste incinerators in Taiwan. *Journal of Environmental Engineering*, 128 (12), 1182–1187.

Chimenos, J.M., Fernandes, A.I., Nadal, R. & Espiell, F. (2000) Short-term natural weathering of MSWI bottom ash. *Journal of Hazardous Materials*, B 79, 287–299.

Colt International Licensing Ltd. Available from: www.coltinfo.co.uk/allington-efw.html.

Conner, J.R. (1990) *Chemical Fixation and Solidification of Hazardous Wastes.* New York, NY, Van Nostrand-Reinhard.

de Waart, H.A.A.M. (2009) Amsterdam waste fired power plant, first year operating experience. In: *Proc. of the 17th Annual North American W-to-E Conference, NAWTEC17, May 18–20, 2009, Chantilly, Virginia, USA.*

Fleck, E. (October 2012) *Waste Incineration in 21st century—Energy Efficient and Climate-Friendly Recycling Plant & Pollutant Sink.*

Freyssinet, Ph., Piantone, P., Azaroual, M., Itard, Y., Clozel-Leloup, B., Guyonnet, D. & Baubron, J.C. (2002) Chemical changes and leachate mass balance of municipal solid waste bottom ash submitted to weathering. *Waste Management*, 22, 159–172.

Geysen, D., Van Gerven, T., Vandecasteele, C., Jaspers, M. & Wauters, G. (2003) Immobilization of heavy metals in fly-ashes and flue gas cleaning residues from MSW incineration—A comparison of the immobilization with cement and with silica containing materials. In: *Proc. of WASCON 2003, Progress on the Road to Sustainability, June 4–6, San Sebastian, Spain.* p. 873.

Gillett, Q. (2011) *Operations at RRR EfW Facility*, CIWM Seminar, 24 June 2011.

Gougar, M., Scheetz, B. & Roy, D. (1996) Ettringite and C–S–H Portland cement phases from waste ion immobilization: A review. *Waste Management*, 13, 295–303.

Granatstein, D.L. & Sano, H. (August 2000) *Toshima Incineration Plant, Tokyo, Japan—A Case study.* Available from: www.ieabioenergytask36.org/Publications/.../Toshima_Case_Study.PDF.

Hitachi Zosen Inova. *Riverside/UK Energy from Waste Plant.* Available from: www.hzinova.com/cms/images/stories/.../hzi_ref_riverside-en.pdf.

Jaspers, M. (2002) Treatment of bottom ashes from waste incineration by the wet process: The Indaver case. In: Wauters, G. & Vandecasteele, C. (eds.) *Annual Symposium on the Technical Aspects of Hazardous Waste Incineration, 27–29 November 2002, Leuven, Belgium.*

JFE Technical Report (March 2011) *JFE High-Temperature Gasifying and Direct Melting Furnace System—Operational Results of Fukuyama Recycle Power, New products & Technologies.* JFE Technical Report no.16. Available from: www.jfe-steel.co.jp/en/research/report/016/pdf/016 05.pdf.

Kleis, H. & Dalager, S. (2004) *100 Years of Waste Incineration in Denmark—From Refuse Destruction Plants to High-Technology Energy Works.* Published by Babcock & Wilcox Volund, and Ramboll.

Lischke, G. *Waste Incineration Technologies from Lentjes GmbH.* Available from: www.lsta.lt/files/events/7_lischke.pdf.

Mainz WtE Plant (December 2007). *The Mainz Waste-to-Energy Plant.* Mainz, Germany, Entsorgungsgesellschaft Mainz mbH. Available from: http://www.mhkw-mainz.de/filead min/downloads/EGM_infobro3_eng_20.pdf.

Martyn, H. (2011) *Development of RRR EfW Facility.* CIWM Waste Seminar, 24 June 2011.

Meima, J.A. & Comans, R.N.J. (1999) The leaching of trace elements from municipal solid waste incinerator bottom ash at different stages of weathering. *Applied Geochemistry,* 14, 159–171.

Nagayama, S. *High Energy Efficiency Thermal WtE Plant for MSW Recycling—JFE High-Temperature Gasifying and Direct Melting Furnace.* Available from: www.iswa.org/uploads/tx_iswaknowledgebase/Nagayama.pdf.

OVAM (2001) *Inventarisatie van de afvalverbrandingssector in Vlaanderen. (Inventory of the Waste Incineration Sector in Flanders) D/2001/5024/7.* Mechelen, Belgium.

Peck, P. Available from: www.metso.com.

Reno Nord. *Waste-to-Energy Plant.* B&W Volund, Plant Fact sheet, Sietse Agema. (2016) Personal communication, Feb. 2016.

Sietse Agema. (2016) Personal communication, Feb. 2016.

Themelis, N.J. & Mussche, C. (2013) Municipal solid waste management and Waste-to-Energy in the United States, China and Japan. In: *2nd International Academic Symposium on Enhanced Landfill Mining, Houthalen-Helchteren, 14–16, October 2013.*

Tolvik (2011) *Waste Market Watch—EfW Operational Performance.*

Unda, J. (2009) *Steps Forward in Energy: Upgrading of WtE Plants.* Presentation at 'Middle East Waste Summit', Dubai, May 27, 2009. Available from: http://www.wtert.com.br/home2010/arquivo/publicacoes/usina_de_zabalgarbi_espanha-ciclo_combinado_gas_natural_RSU.pdf.

Vandecasteele, C., Wauters, G., Arickx, S., Jaspers, M. & Van Gerven, T. (2007) Integrated municipal solid waste treatment using a grate furnace incinerator: The Indaver case. *Waste Management,* 27, 1366–1375.

Van Gerven, T., Geysen, D., Stoffels, L., Jaspers, M., Wauters, G. & Vandecasteele, C. (2005a) Management of incinerator residues in Flanders (Belgium) and in neighbouring countries—A comparison. *Waste Management,* 25, 75–87.

Van Gerven, T., Van Keer, E., Arickx, S., Jaspers, M., Wauters, G. & Vandecasteele, C. (2005b) Carbonation of MSWI-bottom ash to decrease heavy metal leaching, in view of recycling. *Waste Management,* 25, 291–300.

Waste-to-energy-Babcock & Wilcox Volund. Available from: www.volund.dk/References_and_cases/Waste_to_energy_solutions.

WSP (January 2013) *Review of the State-of-the-Art Waste to Energy Technologies, Stage 2—Case Studies.* London, UK, WSP Environmental Ltd. Kevin Whiting, Sr. Tech. Director.

Zabalgarbi website: www.zabalgarbi.com/en/the-zabalgarbi-or-sener-system.html.

Combustion control systems

The waste incineration process is sensitive to the complex chemical processes and inconsistent fuel qualities of the waste such as moisture content and calorific value which are not available online. As a result, the necessityfor reliable control techniques has increased significantly since 2005, especially for waste combustion (Baxter and El Asri 2004, Bardi and Astolfi 2010).

MPC (Model Predictive Control) is an advanced control strategy implemented on a first principle model (Leskens et al. 2005, 2008). This technique is used to optimize the control signals based on two or more inputs/measurements, and to use the system constraints at the same time. In addition to maintaining the variations minimum in the steam production as well as oxygen concentration from their respective setpoint values, the MPC has to restrict the fluctuations in the manipulated variables (i.e. variance). Furthermore, the MPC is submitted to the following constraints: (a) upper and lower bounds on the manipulated variables, and (b) a constant sum for the primary and secondary air flow (Andreas 2004). Andreas (2004) studied the implementation of two different control configurations (feed back and the feed forward) using MPC, to overcome the effect of the heavy disturbances acting on the combustion process, in particular, the caloric value of the waste which largely influences the steam production and the oxygen concentration. The basic principle of the *feedback* configuration is that it compensates for a disturbance only after the controlled variables (cv's), i.e., the steam production and the oxygen concentration, have deviated from their setpoints (Figure A1.1).

The linear MPC (LMPC) model of the plant is generally used because of its computational advantage. The objective of the *feed forward* scheme is to measure the main disturbance, the caloric values of the waste, and compensate for them before the cv's deviates from setpoints. The simulation studies on these two control configurations have shown that the effect of the disturbance on the outputs can be reduced significantly by both the configuration. It is also seen that the feedback control gives a better control performance compared to the feed forward case especially for the steam production. However, a linear model is inadequate to capture all control relevant dynamics, as the large amount of the disturbances imposes the manipulated variables to run over such a large operating range. Leskins et al. (2008) utilized a nonlinear model (NMPC) to overcome both the disadvantages and deliver a better combustion disturbance rejection performance.

(ii) A fuzzy logic control is another strategy to ensure efficient controlled combustion of the municipal waste (e.g., Krause et al. 1994, Müller et al. 1998, Chen 1995,

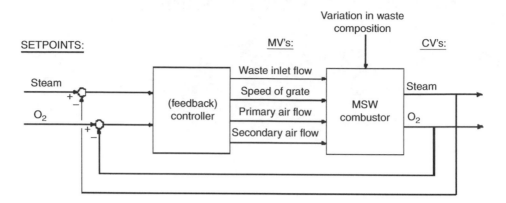

Figure A1.1 The MSW combustion plant and its feedback control system (MPC) (source: Andreas, Arenst 2004).

Gierend and Born 2000, Görner 2001, Chen et al. 2002). In a waste incinerator, the mass flow, calorific value and the mechanical properties of waste are largely inconsistent. For example, the average calorific value can be calculated by an overall balance over the plant, considering the waste as thermal input and the live steam as output. Also, due to the fluctuating quality of waste, the flue gas emissions, carbon burnout in the ash and so on must be controlled within a narrow range. In systems engineering, such boundary conditions are described as fuzzy. The fuzzy logic first proposed by Zadeh (1965a, 1965b) can be applied to such systems very successfully. The fuzzy control system provides an innovative solution, wherein the knowledge and experience of system operators are applied in the system control. i.e., fuzzy controls contain the collected knowledge of the system personnel, with whose help the combustion process in the waste combustion system is controlled. The fuzzy logic-controlled combustion system has potential for most significant cost reduction as well as for large increase in efficiency. Furthermore, waste heat values can be increased while abating pollutant generation. The decrease of pollutants, in turn, prevents corrosion of system components reducing frequent breakdowns and the extent of overhauls resulting from corrosion.

Gierend and Born (2000) in Germany successfully applied this concept to massburn incineration of municipal waste by defining upto 5000 rules obtained by interviewing the operating staff. Chen (1995) discusses how rule-base fuzzy logic control algorithms (Zadeh 1968) can be used to restrict the fluctuation of the steam flow rate and achieve more complete combustion of the waste by suitably adjusting the grate rotating rates. The extensive simulations by Chen (1995) based on the data and information from the Ulu Pandan Waste incineration Plant in Singapore, show the effectiveness of the rule-base fuzzy logic controller, and indicate that the proposed control algorithm has potential of about 10% increase on the capacity of waste processing and electricity generation.

Artificial neural networks

Neural nets can learn the behavior of a Plant by looking on the process inputs and outputs. The transfer function can be approximated to some extent by this training process without introducing special knowledge on the process (Kinnebrock 1992, Krause et al. 1994, Muhlhaus and Gorner 1999, Chen et al. 2002). After the training period, the neural net has the same behavior as the real process. This feature can be utilized for observer control purpose. By inverting the model, an optimization is possible to ask the model, 'the input varibles to a special combination of output variables (Gorner 2001).

Muhlhaus (2000) deduced the complete set of rules (the real process behavior) by training the neural net with measured input and output varibales in the case of a MSW incinerator.

Commercial ACC system: Babcock & Walcox Volund is one of the developers of commercial ACC systems. Their main objective to maintain a uniform flue gas temperature profile, thus reducing the size and number of high temperature regions and to control the primary combustion air distribution and ratio of primary to secondary combustion air flow. In addition, the system controls adjust the position of the main combustion zone to meet variations in the waste heating value. The system sensors are IR cameras. The signals are digitized and analyzed in a neural network-based control system with feed forward signal to the Vølund Control System. The benefits of using the ACC system are: Increased annual waste throughput, Improved steam production with more constant production rate, Reduced stress on the turbine, less maintenance and stops, Reduction in the use of auxilary fuels, Increased lifetime of boiler & refractory through a more constant thermal exposure of the plant components, Reduction of excess air, Reduced emissions, Optimal quality of ashes through systematically controlled burnout, Increased thermal effiency, and Increased availability.

REFERENCES

Andreas, A. (2004) The Comparison of the Feed-back and the Feed-forward Control Configuration in the MSW Combustion Plant. *Prosiding Seminar Nasional Rekayasa Krimia Dan Proses, 2004.* Available from: 118.96.137.51:888/bahanajar/download/.../Solid%20Waste.doc

Babcock & Wilcox Volund: 21st century Advanced Concept for Waste-fired Power plants.

Baxter, D. & El Asri, R. (2004) Process control in municipal solid waste incinerators: Survey and assessment. *Waste Management & Research,* 22, 177–185.

Chen, D. (1995) Fuzzy logic control of batch-feeding refuse incineration. In: *Uncertainty Modeling and Analysis, Annual Conf. of the North American Fuzzy Information Processing Society; Proc. of ISUMA—NAFIPS '95; 17–19 Sept. 1995.* Available from: http://ieeexplore.ieee.org/xpl/articleDetails.jsp?arnumber=527669.

Chen, W.C., Chang, N.B. & Chang, J.C. (2002) GA-based fuzzy neual controller design for municipal incinerators. *Fuzzt Sets Syst.,* 129, 343–369.

Gierend, Chr & Born, M. (2000) Fuzzy control in Waste Incineration Plants – experiences in several types of Plants. *ACHEMA Fair, Frankfurt/M, Germany, 22–27 June 2000.*

Gorner, K. (2001) Waste Incineration – State-of-the-Art and New Developments. *IFRF 13th Members Conference, Noordwijkerhout, The Netherlands, 15–18 May 2001.*

Kinnerbrock, W. (1992) Neuronale Netz, Oldenburg Verlag, Munchen Wien.

Krause, B., von Altrock, K., Limper, K. & Schaefers, W. (1994) A neuro fuzzy adaptive control strategy for refuse incineration plants. *Fuzzy Sets Syst.,* 63 (3), 329–338.

Leskens, M., van Kessel, L.B.M. & Bosgra, O.H. (2005) Model predictive control as a tool for improving the process operation of MSW combustion plants. *Waste Management,* 25, 788–798.

Leskens, M., van der Linden, R.J.P., van Kessel, L.B.M., Bosgra, O.H. & Van den Hof, P.M.J. (2008) Nonlinear model predictive control of municipal solid waste combustion Plants. In *Intl. Workshop on Assessment and Future Directions of NMPC, Pavia, Italy, September 5–9, 2008*. Available from: www.dcsc.tudelft.nl/.../Paperfiles/Leskens&etal_NMPC08_Pavia.pdf.

Leskins, M., van Kessel, L.B.M., Van den Hof, P.M.J. & Bosgra, O.H. *Model Predictive Control of MSW Plants*. Available from: www.dcsc.tudelft.nl/Research/old/project_ml_pvdh_ob.html.

Muhlhaus, R. & Gorner, K. (1999) Feurrungsanalyse und optimierung mit Neuronalen Netzn. *19th German Flame Days, Dresden, Germany, Proceedings, 14/15.9.1999*.

Muhlhaus, R. (2000) Internal report, Lehrstuhl fur Umweltverfahrenstechnik und Analagentechnik, LUAT; taken from Gorner 2001.

Müller, B., Keller, H.B. & Kugele, E. (1998) Fuzzy control in thermal waste treatment. In: *Sixth European Congress on Intelligent Techniques and Soft Computing (EUFIT '98)*. Vol. 3, pp. 1497–1501.

Zadeh, L.A. (1965a) Fuzzy sets. *Information and Control*, 8, 338–353.

Zadeh, L.A. (1965b) Fuzzy sets and systems. In: *Proc. Symp. System Theory, Polytechnic Inst. of Brooklyn, New York*. pp. 29–37.

Zadeh, L.A. (1968) Fuzzy algorithms. *Information and Control*, 12, 94–102.

Infra red sensors

Infrared Radiation (IR) sensors are provided in the advanced control systems for image analysis and detection of burn out lines (e.g., Takatsudo et al. 1999, Manca and Rovaglio 2002. DIAS Infrared GmbH publication, Keller et al. 2007, Matthes et al. 2012). In firing processes for the waste or biomass incineration, camera systems are used for different measurement tasks. Cameras in the VIS range are applied for fixed bed or flue gas burnout monitoring in waste incineration plants. IR cameras with a bandpass filter at $3.9\,\mu m$ are used for the measurement of solid bed temperature distribution because flames are almost transparent at this spectral range.

In order to detect the temperature distribution on the fire-bed of the waste combustion facility, the camera is installed on the boiler at the top of the furnace. Possible disturbances on the optical path between camera and the waste bed surface are minimized by the selection of a suitable spectral region. The required information is selected from infrared photographs by special analytical methods. Thus, for example, it can be determined whether the location and size of the main burnout zone is in the desired region or threatens to run out of this area. In this case, variation of air distribution and/or grate or metering control (length of stroke and frequency) can correct the position of the burnout zone and supply combustion air correctly as needed. However, this is not limited to the whole grate. It is also applicable to specific grate zones individually so that thermal asymmetries are largely reduced (Peter Daimer et al.).

The infrared camera, developed by DIAS Infrared GmbH of Germany and others, to control the combustion process has a small pyroelectric array with 128 elements. With this array it is possible to obtain a thermo picture with 8–16 pixels. This number of pixels is sufficient to control combustion processes in waste incineration plants. As the sensor works at ambient temperature, it needs no cooling. For the modulation of the infrared radiation, a simple chopper is used. The housing of the camera is designed to operate the camera under rough industrial conditions without any additional protections. The quality of the measurement of the temperature is comparable to 128 high grade chopper mode pyrometers. An optional operation with water cooling can extend the ambient temperature of the camera up to 100°C. For the real time pattern correction a digital signal processor is used. The maximum picture repetition is 32 pictures per second. Suitable software is designed to control the measurement process. The camera can be controlled and monitored via a PC. To operate the camera over long distance fiber optics for reliable connections are in use (DIAS Publication). The entry point for the infrared radiation is only about 5 mm in diameter and equipped with an

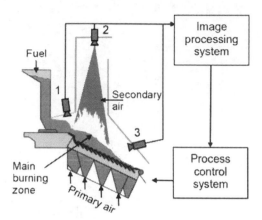

Figure A2.1 Grate firing plant with camera aided combustion control scheme (1, 2 and 3 are cameras) (source: Matthes et al. 2012).

air brush unit. It has an operation range, 650–1500°C, and can resist a temperature stress up to about 1500°C. This infrared camera is reliable and allows a continuous operation possible to enhance the control of the processes in combustion chambers.

Using an image processing system like INSPECT pro control®, new signals can be derived online from the camera images and transmitted to the control system for a process optimization. With this camera based process control CO-peaks could be reduced by 80% and the CO-content by 50%. Additionally the margin of deviation of the steam production could be reduced by 10–30%. As a result the degree of efficiency of the boiler could be increased by 2% (see Figure A2.1) (Keller et al. 2007).

The process control can be further improved by using additional camera based information on the load state of the grate. Matthes et al. (2012) presented a new image processing algorithm which calculates new signals from the IR images indicating a low load state of different regions of the grate. These signals provide a good prediction for drops of the steam production. Utilizing these signals for a better adaptation of the fuel feeding, the deviation of the steam production can be further reduced (Figure A2.1).

REFERENCES

Keller, H.B., Matthes, J., Zipser, S., Schreiner, R., Gohlke, O., Horn, J. & Schonecker, H. (2007) Cameras for Combustion control of highly fluctuating fuel compositions. *VGB PowerTech*, 3/2007.

Manaca, D. & Rovaglio, M. (2002) Infrared thermographic image processing for the operation and control of heterogenous combustion chambers. *Combustion and Flame*, 130 (4), 277–297.

Matthes, J., Waibel, P. & Keller, H.B. (2012) Detection of empty grate regions in firing processes using infrared cameras. *11th International conference on Quantitative Infrared Thermography (QIRT)*.

Takatsudo, Y., Nakamura, N., Ono, H., Mitsuhashi, M. & Kira, M. (1999) Advanced Automatic Combustion Control System for Refuse Incineration Plant. *MHI Technique*. 39 (2), 1999–1995 (in Japanese).

Mathematical modelling of MSW thermal conversion processes

The thermal conversion of municipal solid waste in a WtE plant calls for detailed understanding of the process. The process depends on several input parameters like proximate and ultimate analyses of the waste feed, primary and secondary inlet air velocity, and on the output parameters such as temperature or mass flow rate of conversion products. The variability and mutual dependence of these parameters can be challenging to manage in practice. Another issue of relevance is, how the complex design features of the reactor can be structured to achieve the optimal conversion conditions with minimal pollutant emissions.

Mathematical modelling of the waste combustion, gasification or pyrolysis reactor assures best possible process conditions in order to realise complete thermal conversion of waste with high process effciency.

Mathematical modeling and numerical simulations of fixed beds or fluidized beds in waste incinerators are tricky due to the problems in describing the solid phase (dispersed phase) and its interaction with gas phase (continuous phase). The boundary conditions at the fixed bed surface create problems because of the fuzzy nature of waste properties and the chemical and thermal interaction between the primary combustion air and the waste. For such modelling, advanced computer based engineering tools are used.

CFD approach provides a method to analyze the gas-solid behavior. With ever increasing computer power, CFD modeling has become a useful engineering tool for modeling complex geometry and flow conditions in the incinerators. CFD calculations can give time-dependent information for pressure, temperature, composition and velocity distributions. With such information, the conditions in a reactor can be visualized and parametric studies can be conducted which help in the design process.

Fixed bed modeling have been carried out by different groups (for example, Klasen et al. 1999, Zakaria et al. 2000, Beckmann and Scholz 2000, Yang et al. 2002, 2007, 2008, Ryu et al. 2004, Lim et al. 2001, Haui et al. 2008, Blasaik et al. 2006, Frey et al. 2003, Dong 2000, Kokalji et al. 2005) and some of them have developed their own code.

In a mass burn waste incinerator, the waste undergoes combustion in the bed on the grate and the products of incomplete combustion are destroyed in the gas plenum above the bed. Therefore, it is essential to consider the reaction in both the waste bed and the gas plenum when evaluating the combustion performance using CFD. However, solving the two-phase reacting bed of waste is not straightforward in CFD simulations.

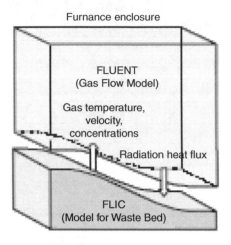

Figure A3.1   Combined simulation strategy (Ryu et al. 2004).

Swithenbank and his group (Ryu et al. 2004), using a new simulation approach that incorporates two models/codes: FLIC for the reacting bed of solid fuels (Yang et al. 2002) and FLUENT for the turbulent reacting flow of the gas plenum has studied the combustion and gas flow of a waste incinerator. The two models were linked through the boundary conditions on the waste bed in such a manner that FLIC provides the gas properties leaving the bed as an inlet condition of the gas flow and FLUENT produces the radiation transferred to the bed (Fig. A3.1). The computed results based on this approach were compared with on-site measurements in the plant by a novel instrument that collected the temperature, oxygen concentration, and motion data while tumbling with the waste particles within the bed.

The measured data were highly fluctuating, but FLIC satisfactorily predicted the overall trend of temperature and oxygen concentration including the upper and lower boundaries of the violent fluctuations. The FLIC/FLUENT combined simulation provided crucial information on the nature of combustion and flow characteristics such as the ignition and burnout points of waste, the rates of each combustion process, and the subsequent gas flow pattern in the combustion chamber (for details, refer to papers by Ryu et al. and Yang et al. from Swithenbank's group).

To reduce the risk arising out of slagging, fouling and corrosion in the waste incinerators, mixing of the secondary air in the furnace is most important because it creates uniform oxygen, temperature and velocity distribution. Klasen and Gorner (2000) developed simple mathematical sub-models for the description of the heterogeneous combustion of the solid waste: The thermal input is defined as the integral of the function "generation of heat" over the grate. The heat release profile along the grate is a function of the axial distance of the waste input and the partition of volatile matter in relation to the sensitive heat at the waste surface. Volatiles emitted from the waste surface are CO and C_xH_y. The gas products CO_2, CO, H_2O and C_xH_y released from the packed bed are calculated in the same way as the heat release. The concentrations of the oxygen are described by an opposite profile over the grate. Gas phase simulations in combination with particle trackings were performed in the complete three dimensional furnace and burnout chamber respectively and the radiational part of a model-MSW incinerator for several cases and furnace geometries with the FLUENT-CFD Code. A higher grid resolution was chosen near the furnace walls and the secondary air nozzles to investigate local temperature and concentration peaks. It was found that an optimization of the incinerator could be achieved by variation of the secondary air injections and the furnace geometry.

Nakamura et al. (2003) developed a mathematical model for time series analysis of continuous variation of MSW using the Monte Carlo method to simulate stochastic combustion, in additionto applying percolation theory in order to simulate transient phenomena. The combustion process on the moving grate is governed by several factors as mentioned above, all of which change continuously. This extreme variability has not been considered in the earlier models of WTE combustion that used average values of the MSW properties. The Monte Carlo stochastic method has been applied to provide a time series description of the continuous variation of solid wastes at the feed end of the traveling grate. The combustion of the solid particles on the grate is simulated using percolation theory. The feed variation and the percolation theory models are combined with the FLIC two-dimensional bed model to project the transient phenomena in the bed, such as the break-up of waste particles and the channeling of combustion air throughout the bed, and their effects on the combustion process. The simulation results of this combustion model with continuous variation of MSW have provided the temperatures and percentage of residual ash by weight. From these values, the percentage volume change from waste to ash has been calculated and the combustion probabilities for each zone have been estimated. The percolation model has shown the mechanism of transient phenomena such as channeling and break-up of solid waste.

Liang and Ma (2010) carried out detailed mathematical modeling and accurate prediction of pollutant emissions. The modeling methods were used for both solid and gaseous phases to simulate the operation of a 450 t/d capacity MSW incinerator to obtain detailed information on the flow and combustion characteristics in the furnace and to predict the amount of pollutant emissions. The predicted data were compared to on-site measurements of gas temperature, gas composition and SNCR de-NO_X system. The major operating conditions considered were grate speed and oxygen concentration. A suitable grate speed ensures complete waste combustion. The predicted data are: volatile release increases with increasing grate speed, and the maximal value is within the range of 700–800 kg/m^2 h; slow grate speeds result in incomplete combustion of fixed carbon; the gas temperature at slow grate speeds is higher due to adequate oxygenation for fixed carbon combustion, and the deviation reaches 200K; NO_X emission decreases, but CO emission and O_2 concentrations increase, and the deviation is 63%, 34% and 35%, respectively. Oxygen-enriched atmospheres promote the destruction of most pollutants due to the high oxygen partial pressure and temperature. The furnace temperature, NO production and CO emission increase as the O_2 concentration increases; the deviation of furnace exit temperature, NO and CO concentrations are 38.26%, 58.43% and 86.67%, respectively. The oxygen concentration is limited to below 35% to prevent excessive CO and NO_x emission without compromising plant performance. The results greatly help to understand the operating characteristics of large-scale MSW combustion plants.

Hussain (2012) has successfully done the simulation of a MSW combustor for different grate furnace designs using the FLIC code and the FLUENT program. Bed combustion was done using FLIC 2.3C; and the CFD modeling of an industrial scale MSW incinerator design was done using FLUENT Ver. 6.2.18. The 2D modeling was based on conservation equations for mass, momentum and energy. The k-turbulence model was employed for turbulent flow modeling. The meshing was done using Gambit 2.2.3. The Generalized Finite Rate model was used to simulate hot flow inside the chamber. The simulations have resulted in predicting the combustion behavior in various grate furnace designs used. The emissions behaviour in various sections of the combustors has been modeled to a reasonable accuracy.

Ismail et al. (2014) have reported a mathematicl model for the drying, volatilization, and combustion phenomena on the packed bed of solid wastes on a moving grate by developing a novel code to investigate combustion process, species concentration, and temperature profile for both gas and solid phases, with less sophistication, and with substantial savings in computational requirements. Gas-solid flow and the heterogeneous chemical reactions were considered. The gas phase turbulence was modeled using k-turbulence model and the particle phase was modeled using kinetic theory of granular flow. The reaction rates of heterogeneous reaction were determined by the Arrhenius-diffusion reaction rate. Flow patterns, gas velocities, particle velocities, composition profiles of gas product and distributions of reaction rates were obtained. Predicted values using the present code were compared with experimental data for validation. The results showed that the predicted exit gas compositions were in good agreement with the experiments.

CFD for reacting multiphase flows is not widely applied mainly because of the difficulty in describing multiphase systems. For example, there is still no agreement on detailed modeling of the interactions between the flowing gas and suspended particles or particle-particle interactions. The incorporation of chemical models in multiphase flow simulations is also a challenging task. Another issue is that multiphase flows are computationally expensive due to a wide range of length and time scales. There is a large amount of published material on the mathematical modeling, and the reader may refer to them for more details.

Bardi and Astolfi (2010) describes a new apprach which focusses on the chemical and thermodynamic aspects of the combustion process, and forms a basis for the development of an advanced controller that uses input/output linearization and extremum seeking (Isidori 1989, Krstic and Wang 2000). This approach has been tested on a model designed and later validated for commercial applications. The results reveal that it is possible to control the waste bed-temperature to specified reference values despite large changes in waste composition. The reference temperature values can be chosen to maximize steam production and at the same time, fulfilling operational conditions. For details, refer to the publication.

Kapitler et al. (2012) used the CFD simulation within ANSYS 13.0 software package and in WORKBENCH 2.0 environment, and employed the supplementary optimisation's methods by using design explorer approach to achieve the target – optimised values of the input and output parameters which have been defined in order to indicate the complete combustion in the furnace. Furthermore, input and output parameters interaction can be found and measured. On this basis, the optimal operating conditions and optimal combustion chamber dimensions are established through decision support systems in order to assure the complete combustion process with minimal emissions on the environment. These requirements could be considered to enhance the operation conditions in the existing furnace or in development or project phase to reduce the costs and time.

Asthana et al. (2010) presents a 2-D steady-state model developed for simulating on-grate municipal solid waste incineration, termed GARBED-ss. Gas–solid reactions, gas flow through the porous waste particle bed, conductive, convective, and radiative heat transfer, drying and pyrolysis of the feed, the emission of volatile species, combustion of the pyrolysis gases, the formation and oxidation of char and its gasification by water vapor and carbon dioxide, and the consequent reduction of the bed volume are described in the bed model. The kinetics of the pyrolysis of cellulosic and non-cellulosic materials were experimentally derived from the actual measurements. The simulation results provide a deep understanding into the various phenomena involved in incineration, for example, the complete consumption of oxygen in a large zone of the bed and a consequent char-gasification zone. The model was successfully validated against experimental measurements in a laboratory batch reactor, using an adapted sister version in a transient regime.

Mathematical modelling is an important aspect in understanding the process and in the design of the WtE plant itself. Though the subject is fairly recent, there are a large number of publications. The reader may find it interesting to go through these papers to get more details on mathematical treatments.

REFERENCES

Anderson, B., Kadirkamanathan, S.R., Chipperfield, V., Sharifi, A.V. & Swithenbank, J. (2005) Multi-objective optimization of operating variables in a waste incineration plant. *Computer & Chemical Engineering.* 29, 1121–1130.

Asthana, A., Menard, Y., Sessiecq, P. & Patisson, F. (2010) Modeling On-Grate MSW Incineration with Experimental Validation in a Batch Incinerator. *Ind. Eng. Chem. Res.*, 49 (16), 7597–7604.

Bardi, S. & Astolfi, A. (December 2010) Modelling and control of a Waste-to-Energy plant- Waste bed temperature regulation. *IEEE Control Systems Magazine*, 27–37.

Beckmann, M. & Scholz, R. (2000) Residence time behavior of solid material in grate systems. In: *5th Europe Conf. Ind. Furnaces and Boilers INFUB, Porto, Portugal,* 11./14./4.2000.

Blasiak, W., Yang, W.H. & Dong, W. (2006) Combustion performance improvement of grate fired furnace using Ecotube system, *Journal of the Energy Institute*, 79 (2), 67–74.

Choi, S., Ryu, C.K. & Shin, D. (1998) A Computational Fluid Dynamics Evaluation of Good Comb. Performance in Waste Incinerators. *J. of the Air and Waste Manag. Assoc.*, 48, 345–351.

Dong, W. (2000) *Design of Advanced Industrial Furnaces Using Numerical Modelling Method.* Doctoral Thesis, Heat and Furnace Technology, Department of Materials Science and Engineering, Royal Institute of Technology, 10044 Stockholm, Sweden.

Frey, H.H., Peters, B., Hunsinger, H. & Vehlow, J. (2003) Characterization of municipal solid waste combustion in a grate furnaces. *Waste Management*, 23, 689–701.

Haui, X.L., Xu, W.L., Qu, Z.Y., Qu, Z.G., Li, Z.G., Zhang, F.P., Xiang, G.M., Zhu, S.Y. & Chen, G. (2008) Numerical simulation of municipal solid waste combustion in a novel two – stage reciprocation incinerator. *Waste Management*, 28, 15–29.

Hussain, A. (2012) CFD Modeling of Grate Furnace Designs for Municipal Solid Waste Combustion. *Asian Transactions on Engineering*, 2 (3), 41–50.

Ismail, T.M., Abd El-Salam, M., El-Kady, M.A. & El-Hagga, S.M. (2014) Three dimensional model of transport and chemical late phenomena on a MSW incinerator. *Intl. Journal of Thermal Sciences*, 77, 139–157.

Kapitler, M., Samec, N. & Kokalj, F. (2012) Operation of Waste-to-Energy-Plant Optimisations by Using Design Exploration. *Advances in Production Engineering & Management*, 7 (2), 101–112; ISSN 1854-6250102. Available at http://dx.doi.org/10.14743/apem2012.2.134.

Klasen, Th. & Görner, K. (1998) Simulation und Optimierung einer Müllverbrennungsanlage. *VDI-GET Fachtagung "Modellierung und Simulation von Dampferzeugern und Feuerungen", Braunschweig, 1998*.

Klasen, T. & Gorner, K. (1999) Numerical calculation and optimization of a large municipal waste incinerator plant. *Second Intl. Symposium on incineration and flue gas treatment technologies, Sheffield University, UK*.

Klasen, T. & Görner, K. (2000) CFD for the prediction of problem areas regarding slagging, fouling and corrosion inside an incinerator. *Proc. 5th European Conference on Industrial Furnaces and Boilers (INFUB), Porto, 11–14 April 2000*.

Klasen, T., Görner, K. & Kümmel, J. Numerische Berechung und Optimierung der MVA Bonn, VDI-Berichte 1492, 19. Deutscher Flammentag, Dresden, 331–336.

Kokalj, F., Samec, N. & Škerget, L. (2005) An Analysis of the Combustion Conditions in the Secondary Chamber of a Pilot – Scale Incinerator Based on Computational Fluid Dynamics. *Strojniški vestnik – Journal of Mechanical Engineering*, 51 (6), 280–303.

Kokalj, F. & Samec, N. (2013) Combustion of Municipal Solid Waste for Power Production, Chapter 9, in 'Advances in Internal Combustion Engines and Fuel Technologie', (ed) Hoon Kiat Ng, ISBN 978-953-51-1048-4, March 20, 2013 under CC BY 3.0 license. © The Author(s), INTECH.

Krüll, F., Kremer, H. & Wirtz, S. Strömungsberechnung zur Bestimmung der Bereiche erhöhter Verschlackungs- und Erosionsgefahr im Kessel einer MVA, VDI-Berichte 1492, 19. Deutscher Flammentag, Dresden, 343–348.

Liang, Z. & Ma, X. (2010). Mathematical modeling of MSW combustion and SNCR in a full-scale municipal incinerator and effects of grate speed and oxygen-enriched atmospheres on operating conditions. *Waste Manag.*, 30 (12), 2520–29; doi: 10.1016/j.wasman.2010.05.006.

Lim, C.N., Goh, Y.R., Nasserzadeh, V., Swithenbank, J. & Riccius, O. (2001) The modelling of solid mixing in municipal waste incinerators. *Powder Technology*, 114, 89–95.

Nakamura, M., Zhang, H., Millrath, K. & Themelis, N.J. (2003) Modeling of Waste-to-Energy Combustion with Continuous Variation of the Solid Waste Fuel. *Proc. of IMECE'03; ASME Intl. Mech. Engg. Congress & Exposition, Washington, D.C. November 16–21, 2003*.

Nasserzadeh, V., Swithenbank, J., Scott, D. & Jones, B. (1991) Design Optimization of a large MSW Incinerator. *Waste Management*, 11, 249–261.

Ravichandran, M. & Gouldin, F.C. (1992) Numerical Simulation of Incinerator Overfire Mixing. *Combustion Science and Technology*, New York, 85, pp. 165–185.

Ryu, C., Yang, Y.-B., Nasserzadeh, V. & Swithenbank, J. (2004) Thermal Reaction Modeling of A large MSW Incinerator. *Combustion Science and Technology*, 176 (11), 1891–1907.

Ryu, C., Yang, Y.-B., Yamouchi, H., Nsserzadeh, V. & Swithenbank, J. Integrated FLIC/FLUENT Modelling of Large Scale MSW Incineration Plants. Available from: www.suwic.group. shef.ac.uk/posters/p-csm ScheffieldMSW.pdf.

Ryu, C., Shin, D. & Choi, S. *Bed combustion and gas flow model for MSW Incinerator*. Department of Mechanical Engineering, Korea Advanced Institute of Science and Technology, Yusong-gu, Taejon, Korea, 305–701.

Van Kessel, L.B.M. (2003) *Stochastic Disturbances and Dynamics of Thermal Processes, with Application to Municipal Solid Waste Combustion*. Doctoral Thesis. Eindhoven, The Netherlands, Eindhoven University of Technology.

Won Yang, Hyung-sik Nam & Cangmin Choi (2007) Improvement of operating conditions in waste incineration using engineering tools. *Waste Manag.*, 27, 604–613.

Yang, Y.B., Goh, Y.R., Zakaria, R., Nasserzadeh, V. & Swithenbank, J. (2002) Mathematical modelling of MSW incineration on a travelling bed. *Waste Manag.*, 22 (4), 369–380.

Yang, Y.B. & Swithenbank, J. (2008) Mathematical modelling of particle mixing effect on the combustion of MS wastes in a packet-bed furnace. *Waste Manage.*, 28, 1290–1300.

Yang, Y.B., Sharifi, N. & Swithenbank, J. (2007) Converting moving-grate incineration from combustion to gasification – Numerical simulation of the burning characteristics. *Waste Management*, 27, 645–655.

Yang, Y.B. & Swithenbank, J. (2008) Mathematical modelling of particle mixing effect on the combustion of MS wastes in a packed-bed furnace. *Waste Management*, 28, 1290–1300.

Zakaria, R., Goh, Y., Yang, Y., Lim, C., Goodfellow, J., Chan, K., Reynolds, G., Ward, D., Siddall, R., Naasserzadeh, V. & Swithenbank, J. (2000) Fundamental aspects of Emissions from the burning bed in a MSW Incinerator. *5th European Conf. Ind. Furnaces and Boilers INFUB, Porto, Portugal, 11./14./4.2000.*

For Simulation of MSW Gasification in Fixed bed gasifiers, the reader may refer, for example, to the following references:

Bryden, K.M. & Ragland, K.W. (1996) Numerical modeling of a deep, fixed bed combustor. *Energy & Fuels*, 10, 269–275.

Chen, C., Jin, Y., Yan, J. & Chi, Y. (2010) Simulation of municipal solid waste gasification for syngas production in fixed bed reactors. *J. Zhejiang Univ-Sci A (Appl Phys & Eng)*, 11 (8), 619–628.

Di Blasi, C.D., Branca, C., Sparano, S. & Lamantia, B. (2003) Drying characteristics of wood cylinders for conditions pertinent to fixed-bed countercurrent gasification. *Biomass & Bioenergy*, 25 (1), 45–58.

Gerun, L., Paraschiv, M., Vijeu, R., Bellettre, J., Tazerout, M., Gobel, B. & Henriksen, U. (2008). Numerical investigation of the partial oxidation in a two-stage downdraft gasifier. *Fuel*, 87 (7), 1383–1393.

Gobel, B., Henriksen, U., Jensen, T.K., Qvale, B. & Houbak, N. (2007) The development of a computer model for a fixed bed gasifier and its use for optimization and control. *Bioresource Technology*, 98 (10), 2043–2052.

Jarungthammachote, S. & Dutta, A. (2007). Thermodynamic equilibrium model and second law analysis of a downdraft waste gasifier. *Energy*, 32 (9), 1660–1669.

Kayal, T.K., Chakravarty, M. & Biswas, G.K. (1997) Mathematical modelling of steady state updraft gasification of jute stick particles of definite sizes packed randomly—An analytical approach. *Bioresource Technology*, 60 (2), 131–141.

Panepint, D. & Genon, G. (2011) Solid waste and Biomass Gasification – Fundamental processes and Numerical simulation. *Chemical Engineering Transactions*, 24, 25–30.

Rogel, A. & Aguillon, J. (2006) The 2D Eulerian approach of entrained flow and temperature in a biomass stratified downdraft gasifier. *American Journal of Applied Sciences*, 3, 2068–2075.

Sharma, A.K. (2008) Equilibrium modeling of global reduction reactions for a downdraft (biomass) gasifier. *Energy Conversion and Management*, 49 (4), 832–842.

Syamlal, M. & Bissett, L. (1992) METC Gasifier Advanced Simulation (MGAS) Model, Morgantown Energy Technology Center, Morgantown, WV, 1992.

Tsai, C.Y. (2011) *A Computational Model for Pyrolysis, heat transfer, and Combustion in a fixed bed Gasifier*. Doctoral Dissertation, Department of Mechanical Engineering, University of Michigan.

Zhang, Q. (2011) *Mathematical modeling of municipal solid waste plasma gasification in a fixed-bed melting reactor*. Doctoral Dissertation, Royal Institute of Technology, School of Industrial Engineering and Management, Department of Material Science and Engineering, Division of Energy and Furnace Technology, Stockholm.

Zhang, Q., Dor, L., Yang, W. & Blasiak, W. (2011) Eulerian model for municipal solid waste gasification in a fixed-bed plasma gasification melting reactor. *Energy & Fuels*, 25, 4129–4137.

Zhanga, Q., Dor, L., Biswas, A.K., Yang, W. & Blasiak, W. (2013) Modeling of steam plasma gasification for municipal solid waste. *Fuel Processing Technology*, 106, 546–554.

Economics of thermal technologies

It is rather a complicated task to determine the cost analysis of an EfW plant utilizing a thermal conversion technology. The chosen technology must meet the following criteria: appropriateness of scale for the particular location/region, matching environmental performance with the local/regional standards, compatibility with the region's goals, and capability of producing marketable products.

The project value tree, in addition to the chosen technology, is influenced by factors such as (a) energy (calorific value) and moisture content of the local waste stream which have to be properly estimated; (b) operational experience at required scale of operation which differs from region to region; for example, the experience in Europe/USA vary considerably with that in Japan. The Japanese installations are state-of-the-art considering the technology, operation, cost, benefits and legislative structure. The technology successfully working at one location may need innovation to install at other location to suit to the quality, quantity and other characteristics of the available waste. This involves contingency costs as well as risks depending on the location which the plant developers/investors have to understand; (c) availability and reliability: operational experience under similar conditions is the best measure of reliability of a thermal technology. But this experience must be at commercial scale; the experience at demonstration level cannot be taken as an adequate proof of process reliability. With unsure reliability, the operational problems lead to frequent plant shutdown for servicing thereby reducing the availability of the plant; (d) fuel flexibility over a long term: the technology must accept the variations in the composition, size distribution and heating value of the input waste which is important for the long term operation. Too much of compositional variations require 'pre-processing' which adds to the cost considerably; (e) mass and energy balance of the plant which provide overall costs and incomes (revenues). Typical aspects to be covered in mass and energy balance are: waste quantity at the pretreatment, unit operations and mass quantities for each flow, air flows for thermal treatment, flue gas flows, input and output water flows, additional materials and chemicals required, output residue quantities for each flow, input thermal energy, productivity of heat energy (for heating purposes), and/or electricity; for economic feasibility, the plant should have atleast two incineration lines, and each line should be capable of handling atleast 240 t/day (10 t/hr); (f) environmental performance: the chosen technology must be environmentally-sound. With stringent emissions regulations, the costs of cleaning equipment increase; (g) assessment of green-house gas emissions as defined by IPCC 2006; (h) costs/economics and scale: the goal of EfW is to move from the concept of waste treatment to energy recovery from the waste. Hence, high electrical efficiency will result in high electricity yield getting high revenues from the sales of electricity and/or heat.

The efficiencies and other cost benefits of the three thermal processes are discussed earlier.

The main *economic* costs considered in an EfW installation are (a) Investment costs: equipment costs that cover thermal treatment equipment, pre-processing costs, flue gas cleaning and ash treatment; costs of working capital; financing and insurance costs and contingency costs. The equipment for flue gas cleaning is decided to a great extent by the desired emission quality level (basic emission level or advanced emission level) which influences the investment costs; (b) Operational costs – includes costs of chemicals, materials and auxiliary fuels; costs of energy consumption; costs of personnel; costs of maintenance (machinery and buildings); costs of byproducts and residues, and variable charges; (c) Revenues (income) – gate fees, energy sales (heat and electricity sales).

There are other factors that influence economics. For example, by diverting MSW for energy recovery, the landfill space which is expensive particularly in urban areas is saved. The cost of land thus saved and associated charges for transporting waste to the landfill areas be included in the analysis. The environmental advantages have to be quantified and considered; for example, greenhouse gas emissions that enhance global warming are considerably lowered with the reduction of landfill. By sorting out non-recyclable plastics from landfills and pyrolysing, value-based transport oils and chemicals can be produced that adds to revenues.

As an illustration, the investment costs as a function of annual and daily capacity for a new MSW incineration plant of capacity 300,000 t/year assuming the calorific value of the waste as 9 MJ/kg for the design, the net treatment cost of plant, energy recovery as a function of calorific value of waste, financing of the plant, cost-benefit analysis are discussed in WB report (1999).

Life cycle cost analysis (LCC)

One of the two methods is followed: (a) Estimating costs of implementing and operating each management system over a 20-year life cycle; or (b) Calculating levelized cost of waste management in units of $/ton of MSW, and determining a discounted payback period and internal rate of return for alternative management systems taking into account tipping fee revenues (Hervin 2013). The Solid Waste Handbook by Robinson (1986) describes the methodology to determine LCC of each management system. The summarized formula is

$$\text{LLC } (\$) = \sum_{n=0}^{n=20} \{(C - R) \times [(1 + i)^n/(1 + d)^n]\}$$

Here, C = annual system cost (real $),
   R = annual system revenue (real $),
   i = inflation rate,
   d = discount rate,
   n = the period in years.

Hervin (2013) used this approach for estimating LCC of the gasification system installed in Humboldt Bay Generating Station (HBGS); the details of estimating LCC and comparison with other management systems are discussed by Hervin.

The information provided by the Plant operators on the costs generally refer to part of the total investment only. In the case of gasification and pyrolysis, the available independent information is much less compared to combustion due to a small operational experience at commercial level. Most of the data on these technologies relates to demonstration level operation. As mentioned at different places in the text, the economic information provided by different sources are inconsistent. As such, comparisons of the value schemes for investment in several types of thermal conversion technologies is not realistic.

REFERENCES

Hervin, K. (May 2013) *Feasibility Analysis of Gasification for Energy Recovery from Residual Solid Waste in Humboldt County*. Thesis submitted to Humboldt State University for the Degree of Master of Environmental systems. Energy Technology and Policy.

IPCC (2006) *2006 IPCC Guidelines for National GHG Inventories*. Vol. 5.

ISWA (January 2013) *Alternative Waste Conversion Technologies, White Paper*. Prepared by International Solid Waste Association Working Group on Energy Recovery; Leading authors: Lamers, F., Fleck, E., Pelloni, L. & Kamuk, B.

Robinson, W. (1986) *The Solid Waste Handbook: A Practical Guide*. Hoboken, N.J, USA, John Wiley & Sons. Inc.

WB (1999) *Municipal Solid Waste Incineration*. World Bank Technical Guidance Report. Washington, DC, The World Bank.

Other references cited at relevant places in the Text.

Subject index